# A Teaching Essay on Residual Stresses and Eigenstrains

# A Teaching Essay on Residual Stresses and Eigenstrains

**Alexander M. Korsunsky**
*University of Oxford*

Butterworth-Heinemann
An imprint of Elsevier

Butterworth-Heinemann is an imprint of Elsevier
The Boulevard, Langford Lane, Kidlington, Oxford OX5 1GB, United Kingdom
50 Hampshire Street, 5th Floor, Cambridge, MA 02139, United States

**Notices**
Knowledge and best practice in this field are constantly changing. As new research and experience broaden our
understanding, changes in research methods, professional practices, or medical treatment may become necessary.

Practitioners and researchers must always rely on their own experience and knowledge in evaluating and using any
information, methods, compounds, or experiments described herein. In using such information or methods they
should be mindful of their own safety and the safety of others, including parties for whom they have a professional
responsibility.

To the fullest extent of the law, neither the Publisher nor the authors, contributors, or editors, assume any liability
for any injury and/or damage to persons or property as a matter of products liability, negligence or otherwise, or
from any use or operation of any methods, products, instructions, or ideas contained in the material herein.

**Library of Congress Cataloging-in-Publication Data**
A catalog record for this book is available from the Library of Congress

**British Library Cataloguing-in-Publication Data**
A catalogue record for this book is available from the British Library

ISBN: 978-0-12-810990-8

For information on all Butterworth-Heinemann publications visit our website at
https://www.elsevier.com/books-and-journals

Working together
to grow libraries in
developing countries

www.elsevier.com • www.bookaid.org

*Publisher:* Matthew Deans
*Acquisition Editor:* Christina Gifford
*Editorial Project Manager:* Anna Valutkevich
*Production Project Manager:* Lisa Jones
*Designer:* Matthew Limbert

Typeset by TNQ Books and Journals

# Contents

# Biography

**Professor Alexander Korsunsky** is a world leader in mechanical microscopy and rich tomography of materials systems and structures for optimization of design, durability, and performance. He heads the Multi-Beam Laboratory for Engineering Microscopy in the University of Oxford and the Centre for In Situ Processing Science in the Research Complex at Harwell Oxford. He consults Rolls-Royce plc on matters of residual stress and structural integrity and is Editor-in-Chief of *Materials & Design*.

In the last two decades, Prof. Korsunsky has been the most active proponent of eigenstrain theory for the analysis of inelastic deformation and residual stresses in materials and components. He teaches widely across the world, and each year he gives several keynote and plenary lectures at major international conferences on engineering and materials. The broader context of Prof. Korsunsky's research interests concern improving the understanding of integrity and reliability of engineered and natural structures and systems, from high-performance metallic alloys to polycrystalline ceramics to natural hard tissue, such as human dentin and seashell nacre. He has coauthored books on fracture mechanics and elasticity and has published ∼300 papers in scholarly periodicals on subjects ranging from multimodal microscopy, neutron and synchrotron X-ray analysis, contact mechanics and structural integrity to microcantilever biosensors, size effects, and scaling transitions. Prof. Korsunsky plays a leading role in the development of large-scale research facilities in the United Kingdom and Europe. He has chaired the Science Advisory Committee at Diamond Light Source and is member of UK delegation to ESRF Council. His activities expand the range of applications of large-scale science to problems in real engineering practice.

Support for Prof. Korsunsky's research comes from EPSRC and STFC (major UK Research Councils) and the European Union, Royal Society, Royal Academy of Engineering, Rolls-Royce, Oxford Instruments, Tescan, NRF (South Africa), DFG (Germany), and other international research foundations.

# Preface

*A.S.Pushkin: On the hills of Georgia. 15 May 1829*
*(AMK translation in Appendix B)*

*V. Gavrilin, Tarantella*
*(Ballet version: V. Vasiliev and E. Maksimova in the film "Anyuta", 1982)*

*K. Petrov-Vodkin, Bathing of the red stallion 1912 (oil on canvas, 160 × 186 cm)*
*Tretyakov Gallery, Moscow*

I began working on residual stress analysis in the context of geomechanics and the exploration for coal, gas, and oil. This was back in the late 1980s, during my study and research practice at the Moscow Physico-Technical Institute (PhysTech) under Sergei A. Christianovich, an outstanding academician who headed the Laboratory for the Mechanics of Non-linear Media at the Institute for Problems in Mechanics, my "research base" (placement) at the time. He became interested in the gas contained in coal seams and sudden hazardous outbursts during mining, a fascinating problem that is both multidisciplinary and multiscale in nature. At the atomic scale, it is related to the molecules of natural gas physically adsorbed at the pore surfaces within coal seams. The strength of this attachment is highly sensitive to the confining pressure: when it is reduced as a result of mining, desorption leads to the release of gas. At the next scale-up, further course of the process is governed by the permeability and strength of the rock mass. In coal seams of low permeability, the buildup of pressure may lead to catastrophic blowout of matter and equipment. This example is an instance of a physicochemically nonlinear system that undergoes significant structural change as a function of residual stress. The change may be sudden and catastrophic, as in the case of coal seam gas outburst, or directed and intended, as in the case of shale gas extraction.

The influence of residual stresses on structural integrity is often similarly drastic and can make the difference between failure and safety. Interestingly, the effect may be particularly strong in cases of short-duration or long-term loading, i.e., under extremes of loading time dependence, such as impact or creep-fatigue. This, in turn, can be understood in terms of the particular significance of tensile stress for damage accumulation and fracture within solid materials. The importance of residual stress control during manufacture has been recognized in engineering design practice and manifests itself in a wide range of ingenious technology solutions employed across the spectrum of applications and material systems: metallic alloys, polymers and composites, ceramics and construction materials, thin films and coatings, semiconductors used in electronics applications, and systems such as armor and spacecraft.

In common with many aspects of engineering, development of ideas and methods in the field often proceeded by trial and error, through identifying treatment conditions that have deleterious or positive effects, and optimizing them by means of trial and error. Although the advent of computers meant a ramp change in the access to numerical modeling tools and versatile characterization equipment, the conventional approach seems to have persisted until now. At least two reasons may be identified why this is the case: (1) the challenge of measuring the complex residual stress tensor at the right location, with the required spatial resolution, preferably in a nondestructive fashion, and (2) the difficulty of interpreting the results in way that is compatible with modern simulation tools used for the determination of safe design life of components and assemblies.

I think this book is timely, because the early decades of the 21st century saw the accumulation of insights, theoretical and experimental approaches, and techniques that led to a step change in our understanding of residual stress states, their origins, nature, and evolution under thermal, mechanical, and other loading conditions. It is timely to introduce some pivotal terms that arose in the context of this subject, such as *eigenstrain theory, mechanical microscopy,* and *residual stress engineering,* to describe specific branches in the system of approaches aimed at systematic description, classification, measurement, modeling, and control of residual stresses.

In my opinion, in this context the theory of eigenstrains and their relationship with residual stresses must play a central role. Over the years I have used the opportunities to teach this subject to different cohorts of undergraduate and postgraduate students: at Oxford, at the National University of Singapore, at Ecole Nationale Supérieure d'Ingénieurs de Caen in Normandy, France. Frequently the students approached me for advice on the choice of a good book that would introduce the concepts and methods that I presented to them in my lectures. Although a number of good volumes exist on the subject, including some collections of contributed articles from prominent researchers in the field, I felt that there was not a text available that would combine an exposition of fundamental ideas with some projections onto particular application fields. After a while, my answer came to be that such book is still being written—so eventually I had to make an effort and express my perception of the subject in the form of this script, which I refer to as a *teaching essay*. By that I want to indicate that it does not presume to be complete but nevertheless seeks to encapsulate in one narration what I consider to be the key elements of knowledge that someone entering the subject would wish to have at their disposal. In the modern day, information is becoming ever more readily available through online resources. The ease of searching for keywords means that, on the one hand, a plethora of articles can be found in seconds, but on the other, one can easily get lost in this sea of detailed studies that may obscure the overall picture. I think that the purpose of an essay like this is to pull together into a coherent narrative various *pointers* to specific results, so that the reader can follow the threads that they find important and relevant to their particular purpose.

Even with the various references I include, I do not consider this essay exhaustive or even complete in any way—it is a snapshot of a work in progress. For example, many miniature modeling tools that I have written in Matlab or Mathematica over the years (and that were the source of some of the illustrations provided) always remain in a state of flux—because new ideas emerge and also because software providers keep changing the underlying functions, which makes old versions stop working. My choice in preparing this report has been to make reference to these tools and give an indication of their functionality: for some readers this will be enough to enable them to write their own code, others may choose to get in touch to ask for help or advice, and then may be able to complete writing something that I never got round to finishing yet! Alternatively, questions at the end of some sections may ask for analysis that is best accomplished by putting together some code. I do not presume to aspire to the level of Landau and Lifschitz physics textbook, some exercises in which, if done properly, to this day may give a diligent student enough material to publish an article.

It is logical and inevitable that the narrative in this essay combines and links the theoretical and numerical modeling frameworks with the experimental approaches for residual stress evaluation. It may be argued that the history of technology is a story of man finding ways to augment physical and mental strengths with ever more complex handy tools, from crowbar to computer.[1] For my research group, throughout the years the tight link between modeling and experimentation has become *modus operandi*, perhaps even *raison d'être*! As will become clear to the reader, the cause of this lies in the very definition of stress (including residual stress), which relies on a mind experiment involving severing internal material bonds and replacing them with equivalent distrib-

---

[1] Put shortly, "If experiment is an extension of our hands, then modeling must be an extension of our mind."

uted forces, something we can never hope to achieve practically but may only continue to come up with new, ever more ingenious ways of approximating.[2] For this reason I am adamantly against using the phrase "stress measurement," and much prefer "stress evaluation": some sort of a model is *always* required to convert observable parameters into *stress*, the mental concept that we find so convenient as a vehicle for thinking about the limits of structural integrity and the conditions of failure.

Throughout these years, teaching has been an integral and important part of my activity. My teaching style has evolved, and I found myself progressively devoting more time to trying to educate by drawing similarities, pointing out connections with something already familiar to the listeners, often from other walks of life that may lie as far afield as philosophy, literature, art, music. I also found that during periods of working on a particular aspect of the subject of residual stresses I was accompanied by some particular image or tune, or even a poem. Pointers to these "snippets" appear at the start of each chapter. I hope the reader will be lenient with me in cases when the connection appears tenuous—often it is circumstantial or coincidental, but I hope that those references might prove amusing for those who wish to follow them and wonder about the link and relevance.

I must conclude this Preface with an expression of gratitude to the wonderful people who were by my side on this journey and without whom I would not know how to go on. My wife Tanya is by far the cleverest person I know—she not only taught me a lot but also always supported me with her inexhaustible reserve of patience, strength, wisdom, and generosity. I am indebted to the members of my research group over the years—too many to mention, truly, but all of whom I remember, and care for each and every one, who in their unique and different ways contributed to the development of my understanding and vision, challenged my views, and focused my mind on the connections that I might have overlooked otherwise.

**Alexander M. Korsunsky**
Oxford, January 2017

---

[2] Put shortly, "Experiment is the art of the possible."

# CHAPTER 1

# Introduction and Outline

*C.P. Cavafy: Ithaka. Collected Poems 1992*

*J. Haydn, Piano Concerto no. 11 in D major*
*Igor Levit and Deutsche Kammerphilarmonie Bremen: 17 May 2015*

*Vassily Kandinsky, Black and Violet (oil on canvas, 78 × 100 cm) 1923*
*Private collection*

When a solid is subjected to external loading, mechanical, thermal, or other, it changes its shape and dimensions. These geometrical changes, described quantitatively by displacements and strains, are the subject of *kinematics*. External loads are transmitted within the material via forces acting between elements across imaginary sections and interfaces. These interactions are quantitatively described by stresses (components of force per unit area) and form the subject of the disciplines of *statics* or *dynamics*. The kinematical and dynamical descriptions of solid deformation are linked by the so-called *constitutive law* that establishes the relationship between stresses and strains (and/or their derivatives and integrals) and captures the peculiarity of a particular material's deformation behavior.

If a solid recovers its original shape and dimensions fully after the removal of the external load, then the deformation (and material) is called *elastic*; otherwise, *inelastic*. If the relationship between strains and stresses in an elastic material is described by a simple linear proportion, then the material is called *linear elastic*; otherwise, *nonlinear elastic*. If a material does not recover its original shape, then it is said to have undergone *inelastic* deformation. In specific cases, the terms plastic strain, creep strain, misfit strain, transformation strain, and eigenstrain are used to refer to the *permanent inelastic strain*.

As this is the first time we encounter the term *eigenstrain*, it is worth introducing the associated notation, $\varepsilon^*$, and recalling that the term was introduced by Toshio Mura (1987) in his book, together with "eigenstress" that he defined as "a generic name given to self-equilibrated internal stresses caused by one or several of these eigenstrains in bodies which are free from any other external

**1**

force and surface constraint." He points out that "eigenstress fields are created by the incompatibility of the eigenstrains," and adds that "engineers have used the term 'residual stresses' for the self-equilibrated internal stresses when they remain in materials after fabrication or plastic deformation." Below we flesh out the implications of the aforementioned definitions by discussing the relationships, form of equations, and problem formulations.

Total strain is additive: it is given by the sum of elastic and inelastic parts:

$$\varepsilon_{ij} = e_{ij} + \varepsilon_{ij}^*. \tag{1.1}$$

Total strain must satisfy the *compatibility* condition: it corresponds to a unique continuous deformation field.

After elastic loading and unloading of a stress-free body all strains return to zero, as do the stresses: no *residual stress* arises. Hence, inelastic deformation is a necessary (but not sufficient!) condition for the creation of residual stresses. Residual stresses arise as a body's response to permanent inelastic strains introduced during the loading–unloading cycle. Therefore, stresses exist only in association with elastic strains: there is no such thing as plastic stress. Stresses must satisfy the condition of static equilibrium.

Permanent inelastic strains (*eigenstrains*) are the source of residual stresses, not the other way round. If eigenstrain is introduced into a body, it responds by developing accommodating elastic strains, which in turn give rise to stresses. The *direct problem* of eigenstrain residual stress analysis is the determination of elastic strains and residual stresses from a given distribution of eigenstrains. The *inverse problem* of eigenstrain residual stress analysis is the problem of finding the eigenstrain distribution that gives rise to the (measured or partially known) distributions of residual elastic strains and/or residual stresses.

Residual stresses *cannot* be measured directly, nor can, indeed, any stress: stress is an imaginary tool used for convenience of description. Instead, other quantities related to residual stresses are measured, usually some displacements or strains, strain increments, or some indirectly related physical quantities, e.g., magnetization. Experimental methods differ in the level of accuracy and detail with which the residual stresses can be assessed: some are only qualitative, whereas others provide highly detailed information that is resolved in terms of spatial location and orientation.

Diffraction provides a uniquely powerful method of evaluating lattice parameters, from which crystal strains can be readily deduced. Diffraction utilizes the simple relationship provided by Bragg's law between the interplanar lattice spacing parameter $d$, the wavelength $\lambda$ (which can be expressed through photon energy $E$ and universal constants $h$ and $c$), and the scattering angle $2\theta$:

$$2d\sin\theta = \lambda = \frac{hc}{E}. \tag{1.2}$$

The lattice spacing $d$ is found experimentally by fixing one of the two parameters $E$ or $\theta$, and varying the other one to find a peak. Consequently, two principal experimental diffraction methods are in use, the angle dispersive (monochromatic, $\lambda =$ const) or energy dispersive (white beam, $2\theta =$ const). Highly detailed information can be extracted by simultaneous interpretation of multiple diffraction peaks, e.g., the strain average across all peaks provides information about the equivalent macroscopic strain, whereas the differences reflect the inhomogeneity of deformation in terms of the interaction, and load and strain transfer between crystallites of different orientation.

Solid objects are very rarely (in fact, almost never) the idealized homogeneous and isotropic continua studied in the classical mechanics of deformable solids. Real materials consist of distinct grains and phases, e.g., polycrystals that are made up of constituent grains whose mechanical properties depend on orientation. Loading applied to such "composites" induces inhomogeneous deformation: some regions experience greater strain or stress than others; some may yield, whereas others remain elastic. As a consequence of this inhomogeneity, after unloading, some *microscopic intergranular* residual stresses arise.

Within individual grains inelastic deformation is often mediated by crystallographic defects, such as *dislocations*. Dislocations appear and move under the applied shear stress; they exert forces upon each other and can organize themselves into more or less stable arrangements, such as persistent slip bands, pile-ups, ladders, and cell-wall structures. Dislocations can escape through the material free surface, leaving behind steps, and pairs of dislocations of opposite sign can annihilate. Defects such as dislocations give rise to residual stresses at the *intragranular* level.

This book contains an introductory course devoted to the analysis of residual stresses and their interpretation and understanding with the help of the concept and modeling framework of *eigenstrain*. The following order of presentation is adopted. After a brief introduction to the subject in this chapter, the continuum mechanics fundamentals for the description of deformation and stresses are given in Chapter 2. Some basic types of inelastic deformation and residual stress states are defined and analyzed: the consideration of constraint and stress balance conditions allows the introduction of the simple (i.e., 0-D, or "point") residual stress states in Chapter 3. In Chapters 4 and 5, the important case of one-dimensional residual stress states is addressed, using the examples of inelastic beam bending and inelastic expansion of a hollow tube under internal pressure.

Chapter 6 is devoted to the eigenstrain theory of residual stress. First, the procedure for the solution of the direct problem of eigenstrain is illustrated using the examples of "shrink-fit" eigenstrain cylinder and eigenstrain sphere. The important Eshelby solution for the eigenstrain ellipsoid is introduced next, followed by the concept of *nuclei of strain* as the sources of residual stress concentrated to a point.

Chapter 7 is devoted to the concept of *dislocation* because of its particular significance in materials science and engineering. After a brief overview of the nature of dislocation and the associated elastic solutions, attention is given to inelastic deformation and residual stress generation by means of dislocation movement (discrete dislocation dynamics). Applications to practically important cases are presented and discussed.

Chapter 8 is devoted to a brief overview of the experimental methods for residual stress evaluation and their interpretation. This is particularly important in view of the fact that direct measurement of residual stress turns out to be impossible almost in all cases, so that the evaluation of residual stress is possible only by adopting some sort of an interpretative modeling approach. The emphasis is placed on the classification of experimental techniques and the discussion of their suitability in different practical situations, their limitations, and the outlook for their development in terms of spatial resolution and reliability.

Owing to the importance of micro- and nanoscale residual stress evaluation methods in the context of current research and technology, the entire Chapter 9 is devoted to this topic. It begins with an overview of microfocus diffraction techniques with X-ray and electron beams and spectroscopic methods, e.g., Raman. The rest of the chapter is devoted to the history of the development and illustration of the use of the Focused Ion Beam—Digital Image Correlation (FIB-DIC) technique, with particular attention paid to the microscale ring-core method. FIB-DIC spatially resolved residual stress profiling is classified into sequential and parallel approaches, and case studies exemplifying the use of these approaches are described.

Chapter 10 contains the description of the inverse eigenstrain theory of residual stress reconstruction. After the inverse problem of residual stress determination is formulated, possible strategies for obtaining the solution are discussed, and the eigenstrain-based approach is advocated. The inversion procedure based on least-squares minimization is presented, and the existence and uniqueness of the approximate solution are demonstrated. The choice of the functional basis for seeking the solution is considered, and the conditions for the convergence of the solution to the correct eigenstrain distribution are discussed.

Chapter 11 is devoted to the methods of modeling the interaction between the residual stress and applied external loading. It introduces the concept of current state modeling, as an alternative to process modeling, and emphasizes the interaction between external loading and residual stresses in tribology (surface contact in fretting), and briefly touches upon the possible applications in fracture mechanics.

Finally, Chapter 12 is devoted to conclusions and the outlook.

CHAPTER 2

# Elastic and Inelastic Deformation and Residual Stress

*Marcel Proust: Sur ce coteau normand etablis ta retraite, 1877*

*Philippe Rombi, Cinq Fois Deux, 2004*

*Edouard Manet: Luncheon on the Grass (oil on canvas, 208 × 265 cm) 1862. Musée d'Orsay, Paris*

When a real material is studied at progressively diminishing scales, it is usually found to exhibit a variation in properties from point to point. The spatial variation of properties is called *structure*. The term is often used with reference to a certain scale, e.g., atomic structure, nanoscale structure, microstructure. The mechanical response of material to loading depends on its structure, and a large section of materials science is devoted to the analysis of this dependence. Nevertheless, some fundamental aspects of mechanical behavior can be understood very well if the material is considered to be structureless, a solid continuum. Historically, the development of continuum mechanics is closely linked to the growth of natural science at large (e.g., tensor theory, numerical methods, etc.)

Solids respond to applied external loads by developing internal forces. If an imaginary section through a solid is considered, the components of internal force acting on a unit elemental area are called stresses. Under the action of stresses solids deform, so that the distances between points change. However, provided the stresses are sufficiently small, the solid recovers its original shape and volume once the load is removed. This type of behavior is called *elastic*. Continuum elasticity considers the consequence of atomic interactions in solids, but disregards their nature. Inelastic behavior is manifested in residual deformation persisting after load removal, and often gives rise to residual stress.

Under very low loads, deformation is found to be proportional to stress, the case of *linear continuum elasticity*. In the following discussion, the basic foundations of stress and strain analysis are laid out in the infinitesimal limit. Then simple forms of the elastic equations for isotropic bodies are introduced in terms of Lamé constants, as well as Young's modulus and Poisson's ratio.

## CONTENTS

5

A Teaching Essay on Residual Stresses and Eigenstrains. http://dx.doi.org/10.1016/B978-0-12-810990-8.00002-1

## 2.1 DEFORMATION AND STRAIN

Each point in a solid continuum can be referred to by its initial position with respect to a Cartesian coordinate system, $x = (x_1, x_2, x_3)$. During deformation, the displacement of each point can be described by the vector $u = (u_1, u_2, u_3)$. The displacement $u$ is therefore a vector function of the position vector $x$. At a point removed from $x$ by $dx = (dx_1, dx_2, dx_3)$, the displacement vector differs by $du = (du_1, du_2, du_3)$. The initial distance between any two closely positioned points given by

$$dl = \sqrt{dx_1^2 + dx_2^2 + dx_3^2},$$

after deformation changes to

$$dl' = \sqrt{(dx_1 + du_1)^2 + (dx_2 + du_2)^2 + (dx_3 + du_3)^2}.$$

The displacement differences $du_i$ can be written in terms of displacement gradients at point $x_j$ (Fig. 2.1). Introducing the expression $\varepsilon_{ij} = \frac{1}{2}\left(\frac{\partial u_i}{\partial x_j} + \frac{\partial u_j}{\partial x_i} + \frac{\partial u_i}{\partial x_j}\frac{\partial u_j}{\partial x_i}\right)$, the elemental length can be expressed as

$$dl'^2 = dl^2 + 2\varepsilon_{ij}dx_i dx_j + o\left(|dx|^2\right).$$

Here, $o\left(|dx|^2\right)$ denotes the term that decays more rapidly than $|dx|^2$ as $dx$ vanishes. The terms discarded here contain strain gradients, i.e., higher order spatial derivatives of displacements. Although ignored in the most fundamental

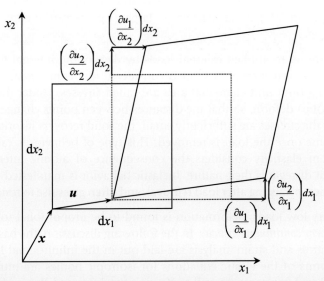

**FIGURE 2.1**

Illustration of strain. *From Korsunsky, A.M., 2001. (4.1.1a) Elastic Behavior of Materials: Continuum Aspects, contributed to Elsevier Encyclopedia of Materials Science and Technology.*

formulation of linear elasticity, these terms can be taken into account in formulating a higher order linear dependence of stress upon strain. They give rise to characteristic length scales related to the nature of deformation.

Following Cauchy, the last term in $\varepsilon_{ij}$ is usually neglected for small deformations, so that

$$\varepsilon_{ij} = \frac{1}{2}\left(\frac{\partial u_i}{\partial x_j} + \frac{\partial u_j}{\partial x_i}\right). \tag{2.1}$$

The terms $\varepsilon_{ij}$ defined according to Eq. (2.1) are called strains and form a symmetric second rank tensor called the *strain tensor*. Strains can be separated into two groups. If $i = j$, then e.g., $\varepsilon_{11} = \partial u_1/\partial x_1$, $\varepsilon_{22}$, $\varepsilon_{33}$ are called *direct*, or *normal, strains* and describe elongation or contraction. Otherwise, $i \neq j$, and $\varepsilon_{12} = 1/2(\partial u_1/\partial x_2 + \partial u_2/\partial x_1)$, $\varepsilon_{23}$, $\varepsilon_{31}$ are called *shear strains*, and describe the change in the angle between two initially straight lines along the coordinate axes (Fig. 2.1).

An alternative orthonormal system of axes can be found by rotation with respect to which shear strains vanish. The remaining normal strains with respect to these so-called *principal axes* are the *principal strains*. A volume originally occupying an elemental cube with face normals along the principal axes deforms into a parallelepiped. The relative volume change (the *dilatation*) is described by the sum of normal strains, $\Delta = \varepsilon_{11} + \varepsilon_{22} + \varepsilon_{33}$. The sum of normal strains is invariant, i.e., remains the same with respect to any coordinate system.

For small deformations, uniform strains are adequately approximated by the ratios of elongations $\Delta l_i$ to the original gauge lengths $l_j$, e.g., $\varepsilon_{11} = \Delta l_1/l_1$, and are referred to as *engineering strains*. For larger distortions, *true* or *logarithmic strains* $\varepsilon_1^T = \ln(1 + \Delta l_1/l_1)$ are often used.

For a given strain field to be suitable for the description of continuum deformation, additional conditions must be imposed. These conditions that ensure the existence of a displacement field from which the strain field can be derived using the above formulas are known as Saint-Venant's compatibility equations, and can be written in the forms similar to the following:

$$\frac{\partial \varepsilon_{ij}}{\partial x_k \partial x_l} + \frac{\partial \varepsilon_{kl}}{\partial x_i \partial x_j} - \frac{\partial \varepsilon_{il}}{\partial x_j \partial x_k} - \frac{\partial \varepsilon_{jk}}{\partial x_i \partial x_l} = 0.$$

## 2.2 STRESS

A general element of cross-sectional area $dA$ possesses an orientation described by a unit normal $n=(n_1, n_2, n_3)$, and transmits an internal force $dF = (dF_1, dF_2, dF_3)$. The force has magnitude $dF = sdA$, and can be resolved into the normal $\sigma dA$ and shear $\tau dA$ components (Fig. 2.2). In particular, for each rectangular element $dA_i$ with normal in the direction $Ox_i$, the internal force can be resolved

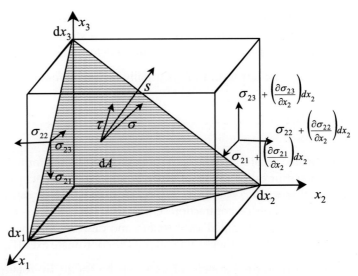

**FIGURE 2.2**

Illustration of stress. *From Korsunsky, A.M., 2001. (4.1.1a) Elastic behavior of materials: Continuum aspects, contributed to Elsevier Encyclopedia of Materials Science and Technology.*

into three Cartesian components written as $dF_j = \sigma_{ij}dA_i$. The terms $\sigma_{ij}$ form a second rank tensor called the *stress tensor*. Components of torque exerted on an elemental volume $dV$ by the surrounding solid can be written as $(\sigma_{23} - \sigma_{32})$ $dV$, $(\sigma_{31} - \sigma_{13})dV$, $(\sigma_{12} - \sigma_{21})dV$. The symmetry of the stress tensor, $\sigma_{ij} = \sigma_{ji}$, follows from the requirement that this expression must vanish in the equilibrium. This assumption holds in the absence of body torque exerted by long-range forces, e.g., the action of a magnetic field on magnetic crystals.

If $i = j$, the terms $\sigma_{11}$, $\sigma_{22}$, $\sigma_{33}$ are called *normal stresses*, and can be classified further into tensile (positive) and compressive (negative). Otherwise, when $i \neq j$, the terms $\sigma_{23}$, $\sigma_{13}$, $\sigma_{12}$ are called *shear stresses*. By coordinate rotation, an alternative orthonormal system of axes (called the *principal axes*) can be found with respect to which shear stresses vanish, and an elemental cube is subjected to normal stresses only (called the *principal stresses*). The sum of normal stresses $\sigma_{11} + \sigma_{22} + \sigma_{33} = -3P$ is *invariant*, i.e., remains the same in any rotated system. It represents the *hydrostatic* component of the stress state, where $P$ denotes pressure (assumed positive when compressive). The stress tensor that remains after subtracting the hydrostatic part from the total stress is known as *deviatoric*.

## 2.3 EQUATIONS OF EQUILIBRIUM

For a solid body to remain in equilibrium, the balance of internal forces acting on any elemental volume within the body must be maintained. By considering a parallelepiped with edges along the coordinate axes (Fig. 2.2), the following equation is obtained:

$$\frac{\partial \sigma_{ij}}{\partial x_j} + B_i = 0, \tag{2.2}$$

where $B_i$ denotes the volume density of long-range forces, e.g., gravity. Deformation of a solid body is quasistatic if inertial forces in this equation can be neglected. When non-quasistatic deformations are considered, as in the case of elastic vibrations and waves, the right-hand side of this equation must contain the term $\rho \ddot{u}_i$.

If the elemental volume considered lies at the boundary, the internal stresses must be balanced by the surface tractions $f_i$, so that

$$\sigma_{ij} n_j = f_i, \tag{2.3}$$

where $n_i$ denotes the components of the outward surface normal.

## 2.4 FORMULATION AND SOLUTION OF PROBLEMS IN CONTINUUM MECHANICS

The equations of kinematics and statics of continua introduced earlier establish a separate description of distinct aspects of deformation and internal stress. To obtain a well-posed problem in continuum mechanics it is necessary to complete the formulation by introducing the relationship between the quantities related to geometric changes (strains) on the one hand and the quantities related to the transmission of internal forces (stresses) on the other. This connection is provided by the so-called *constitutive equations* that express stresses (or their increments during path-dependent loading) and strains (or their increments). The simplest formulation of the stress–strain relationship is the linear form, or Hooke's law. Most solids display a linear stress–strain behavior for small deformation. Furthermore, more complex constitutive relationships can be formulated starting with the assumption that strain can be split into the elastic and inelastic parts, with the first term being path independent and directly linked to stress, and the second term being dependent on deformation history expressed through so-called internal variables. Therefore, constitutive equations must also describe the evolution of internal variables and define the *partitioning* of total strain into reversible (elastic) and inelastic parts (permanent, thermal, etc.) (Fig. 2.3).

## 2.5 STRAIN ENERGY DENSITY

The increment of strain energy density (i.e., elastic energy per unit volume) in a deforming solid is equal to the work done by the stresses to alter the strains,

Kinematics

Definition of strain: $e_{ij} = \frac{1}{2}\left(\frac{\partial u_i}{\partial x_j} + \frac{\partial u_j}{\partial x_i}\right).$

Compatibility: $\left(\text{curl}\left(\text{curl } e_{ij}\right)^T\right)^T = 0.$

Boundary conditions: $u_i|_{d\Omega} = u_i^*|_{d\Omega}.$

Statics

Stress equilibrium:　　$\sigma_{ij,j} + f_i = 0.$

Moment equilibrium:　　$\sigma_{ij} = \sigma_{ji}.$

Boundary conditions:　　$\sigma_{ij}n_j = t_i^*.$

Constitutive equations

Strain additivity:　　　$\varepsilon_{ij} = e_{ij} + \varepsilon_{ij}^*.$

Hooke's law:　　　　　$\sigma_{ij} = C_{ijkl}e_{kl}.$

Internal variables:　　$d\varepsilon_{ij}^* = F\left(e_{ij}, \varepsilon_{ij}^*, \sigma_{kl}, \alpha, \beta, \gamma, ...\right)d\lambda$

$d\alpha = F\left(e_{ij}, \varepsilon_{ij}^*, \sigma_{kl}, \alpha, \beta, \gamma, ...\right)d\lambda$

$\cdots$

**FIGURE 2.3**
Schematic illustration of continuum mechanics problem formulation.

$dU = \sigma_{ij}d\varepsilon_{ij}$. The relationship between stresses and strains can therefore be expressed in terms of the strain energy density as

$$\sigma_{ij} = \frac{\partial U}{\partial \varepsilon_{ij}}. \tag{2.4}$$

Assuming the deformation considered is small, the energy density $U$ may be expanded into Taylor series in terms of strains with respect to the reference state. If this initial undeformed state of the solid is assumed to be stress-free, then at $\varepsilon_{ij} = 0$ there must be $\sigma_{ij} = 0$. Hence, the linear terms in the expansion must vanish, and the expression for the strain energy density should contain expressions quadratic in terms of strains,

$$U = U_0 + \sum_{ijkl} \frac{1}{2} C_{ijkl}\varepsilon_{ij}\varepsilon_{kl}. \tag{2.5}$$

Alongside *deformation* terms, the total internal energy of a solid contains *entropy* terms. These terms are neglected in the present analysis, although they may make a substantial contribution to the elastic effects in some cases, e.g., in the case of configurational entropy of long molecules in polymeric solids.

Eqs. (2.4) and (2.5) give rise to a linear relationship between stresses and strains,

$$\sigma_{ij} = C_{ijkl}\varepsilon_{kl}. \tag{2.6}$$

Note that in the abovementioned expression the sum sign was omitted, by using the Einstein summation convention that implies summing over each pair of repeated indices.

Higher order terms may appear in the equation and give rise to nonlinear stress–strain dependence. A mathematical treatment of nonlinear elastic deformations can be found in Ogden (1984).

Eq. (2.6) states the *generalized Hooke's law* for a linear elastic solid continuum. The terms $C_{ijkl}$ are called *stiffness coefficients* and form a fourth rank tensor. For strains to be found in terms of stresses, the aforementioned expression may be inverted,

$$\varepsilon_{ij} = s_{ijkl}\sigma_{kl}. \tag{2.7}$$

The terms $s_{ijkl}$ are called *compliance coefficients* and form a fourth rank tensor equal to the inverse of tensor $C_{ijkl}$.

The tensorial property of $\varepsilon_{ij}$, $C_{ijkl}$ (and $\sigma_{ij}$, $s_{ijkl}$) is manifested in their behavior under the rotation of coordinate system. If a new Cartesian coordinate system is defined by the equations

$$x'_i = t_{ij}x_j, \tag{2.8}$$

then the tensor components in the new system are related to the original ones by

$$\varepsilon'_{pq} = t_{pi}t_{qj}\varepsilon_{ij}, \quad C'_{pqrs} = t_{pi}t_{qj}t_{rk}t_{sl}C_{ijkl}. \tag{2.9}$$

## 2.6 CONTRACTED NOTATION

Tensors $\varepsilon_{ij}$ and $\sigma_{ij}$ are symmetric with respect to the interchange of their indices, which must be reflected in the minor indicial symmetries of $C_{ijkl}$ and $s_{ijkl}$: $i \leftrightarrow j$, $k \leftrightarrow l$. Eqs. (2.4), (2.6), and (2.7) shows that $C_{ijkl}$ and $s_{ijkl}$ must remain unaltered if the first pair of indices is interchanged with the last (major indicial symmetry). To take advantage of these symmetries, contracted notation can be introduced that uses six-component vectors for stress and strain, defined as follows:

$$
\begin{pmatrix} \sigma_1 \\ \sigma_2 \\ \sigma_3 \\ \sigma_4 \\ \sigma_5 \\ \sigma_6 \end{pmatrix} = \begin{pmatrix} \sigma_{11} \\ \sigma_{22} \\ \sigma_{33} \\ \sigma_{23} \\ \sigma_{31} \\ \sigma_{12} \end{pmatrix}, \quad
\begin{pmatrix} \varepsilon_1 \\ \varepsilon_2 \\ \varepsilon_3 \\ \varepsilon_4 \\ \varepsilon_5 \\ \varepsilon_6 \end{pmatrix} = \begin{pmatrix} \varepsilon_{11} \\ \varepsilon_{22} \\ \varepsilon_{33} \\ 2\varepsilon_{23} \\ 2\varepsilon_{31} \\ 2\varepsilon_{12} \end{pmatrix}. \tag{2.10}
$$

The contraction of each pair of indices into one is carried out according to the rule:

$$11 \rightarrow 1, \quad 22 \rightarrow 2, \quad 33 \rightarrow 3, \quad 23, \quad 32 \rightarrow 4, \quad 31, \quad 13 \rightarrow 5, \quad 12, \quad 21 \rightarrow 6. \qquad (2.11)$$

The $6 \times 6$ stiffness matrix is introduced using the above index transformation rule as follows:

$$C_{\alpha\beta} = C_{ijkl}, \quad (\alpha, \beta = 1...6, \quad i, j, k, l = 1...3), \qquad (2.12)$$

so that Eq. (2.6) becomes

$$\sigma_\alpha = C_{\alpha\beta}\varepsilon_\beta, \quad \text{where} \quad \alpha, \beta = 1...6. \qquad (2.13)$$

It is desirable to obtain an expression for the 6-strains in terms of 6-stresses and the compliance matrix in the same form as above,

$$\varepsilon_\alpha = s_{\alpha\beta}\sigma_\beta. \qquad (2.14)$$

The change of tensor rank, however, imposes the need for the introduction of factors 2 and 4 as follows:

$$s_{\alpha\beta} = s_{ijkl}, \quad \text{when both } \alpha \text{ and } \beta \text{ are } 1, 2 \text{ or } 3;$$

$$s_{\alpha\beta} = 2s_{ijkl}, \quad \text{when either } \alpha \text{ or } \beta \text{ are } 4, 5 \text{ or } 6; \qquad (2.15)$$

$$s_{\alpha\beta} = 4s_{ijkl}, \quad \text{when both } \alpha \text{ or } \beta \text{ are } 4, 5 \text{ or } 6.$$

In the contracted notation, the number of stiffness or compliance components used to specify the elastic material properties is reduced from $3 \times 3 \times 3 \times 3 = 81$ to $6 + 5 + 4 + 3 + 2 + 1 = 21$ terms of $6 \times 6 = 36$. This is the maximum number of independent elasticity coefficients needed to describe a general anisotropic continuum solid. The elastic symmetry characteristic of single crystal systems reduces this number further. Many polycrystalline solids display directional dependence of elastic properties because of the preferred orientation of their constituents. These systems can be described as *anisotropic elastic continua*.

## 2.7   ELASTIC ISOTROPY

A case of particular practical importance concerns solid bodies that develop the same strain independently of the direction in which the stress is applied. These are called elastically isotropic. Although few single crystals approach this type of behavior, glasses and amorphous solids may be treated as macroscopically isotropic, as can be polycrystals possessing no preferred orientation.

The strain energy density (Eq. 2.5) of isotropic solids must depend on two rotation-invariant second-order scalars composed of the strain components and can be written as follows:

$$U = U_0 + \frac{\lambda}{2}\varepsilon_{ii}^2 + \mu\varepsilon_{ij}^2. \tag{2.16}$$

Isotropic elasticity is fully described by two parameters $\lambda$ and $\mu$, known as *Lamé's constants*. Hooke's law for the isotropic body is obtained by differentiation according to Eq. (2.4):

$$\sigma_{ij} = 2\mu\varepsilon_{ij} + \lambda\delta_{ij}\varepsilon_{mm}. \tag{2.17}$$

Here Kronecker's delta symbol $\delta_{ij} = 1$ if $i = j$, and $\delta_{ij} = 0$ otherwise. Inverting Eq. (2.17) gives

$$\varepsilon_{ij} = \frac{1}{2\mu}\left(\sigma_{ij} - \frac{\lambda}{3\lambda + 2\mu}\delta_{ij}\sigma_{mm}\right). \tag{2.18}$$

## 2.8  ELASTIC CONSTANTS

In the engineering practice, elastic properties of materials are often quoted in terms of Young's modulus and Poisson's ratio. Table 2.1−2.3 lists these parameters for common metallic alloys, ceramics and glasses, and polymers, respectively. Poisson's ratios of ceramics lie in the range 0.1−0.3, whereas for metallic alloys they are close to 0.3. Poisson's ratio of a polymer is not a constant but shows strong dependence on stress, temperature, and time. Values generally lie in the range 0.3−0.5; those appearing in the table are given only as guidelines. Young's modulus shows a higher degree of variability between material classes, associated with the nature of atomic bonding.

**Table 2.1** Elastic Properties of Some Engineering Metallic Alloys

| Engineering Alloys | Young's Modulus, GPa | Poisson's Ratio |
| --- | --- | --- |
| Aluminum | 65−72 | 0.33−0.34 |
| Copper | 100−120 | 0.34−0.35 |
| Magnesium | 45 | 0.3−0.35 |
| Nickel | 200−220 | 0.31 |
| Steels | 200−215 | 0.27−0.29 |
| Titanium | 110−120 | 0.36 |
| Zinc | 105 | 0.35 |

**Table 2.2** Elastic Properties of Some Engineering Ceramics and Glasses

| Engineering Ceramics and Glasses | Young's Modulus, GPa | Poisson's Ratio |
|---|---|---|
| Titanium diboride, $TiB_2$ | 540 | 0.11 |
| Silicon carbide, SiC | 400 | 0.19 |
| Titanium carbide, TiC | 440 | 0.19 |
| Tungsten carbide, WC | 670–710 | 0.24 |
| Silicon nitride, $Si_3N_4$ | 110–325 | 0.22–0.27 |
| Alumina, $Al_2O_3$ | 345–414 | 0.21–0.27 |
| Beryllium oxide, BeO | 300–317 | 0.26–0.34 |
| Zirconia, $ZrO_2$ | 97–207 | 0.32–0.34 |
| Fused silica | 71 | 0.17 |
| Soda-lime glass | 69 | 0.24 |
| Aluminosilicate glass | 88 | 0.25 |
| Borosilicate glass | 63 | 0.20 |
| High-lead glass | 51 | 0.22 |

**Table 2.3** Elastic Properties of Some Engineering Polymers

| Polymers | Young's Modulus, GPa | Poisson's Ratio |
|---|---|---|
| Acrylics | 2.4–3.1 | 0.33–0.39 |
| Epoxies | 2.6–3.1 | 0.33–0.37 |
| Polystyrenes | 3.1 | 0.33 |
| Low-density polyethylene | 0.1–0.3 | 0.45 |
| High-density polyethylene | 0.4–1.4 | 0.34 |
| Polypropylene | 0.5–1.9 | 0.36–0.40 |
| Polytetrafluoroethylene | 0.4–1.6 | 0.40–0.46 |
| Polyurethanes | 0.006–0.4 | 0.49 |

## 2.9 UNIFORM DEFORMATION

The simplest cases of deformation are observed when the strain tensor is uniform throughout the solid body.

*Uniaxial tension.* A slender rod subjected to longitudinal tension develops uniform strains $\varepsilon_{ij}$ and uniform stresses $\sigma_{ij}$. Any section transmits axial stress alone, $\sigma_{11}$, whereas all other stress components are equal to zero. This state of stress is called *uniaxial tension.* From Eq. (2.18), the axial elongation strain $\varepsilon_{11}$ is found as

$$\varepsilon_{11} = \sigma_{11} \frac{(\lambda + \mu)}{\mu(3\lambda + 2\mu)} = \frac{\sigma_{11}}{E}, \quad E = \frac{\mu(3\lambda + 2\mu)}{(\lambda + \mu)}. \tag{2.19}$$

The coefficient $E$ is called *Young's modulus*. The ratio of transverse contraction experienced by the rod to axial elongation is called *Poisson's ratio* and is given by

$$\nu = -\frac{\varepsilon_{22}}{\varepsilon_{11}} = \frac{\lambda}{2(\lambda + \mu)}. \tag{2.20}$$

*Equiaxial compression.* Consider a body subjected to hydrostatic pressure $P$, $\sigma_{ij} = -P\delta_{ij}$. The *bulk modulus K* of an isotropic body is found as the ratio of $P$ to the relative volume decrease, $\Delta = \varepsilon_{ii}$. Summing up Eq. (2.19) for $i = j = 1...3$:

$$\frac{\sigma_{11} + \sigma_{22} + \sigma_{33}}{3(\varepsilon_{11} + \varepsilon_{22} + \varepsilon_{33})} = K = \lambda + \frac{2}{3}\mu = \frac{E}{3(1 - 2\nu)}. \tag{2.21}$$

*Pure shear.* Consider uniform pure shear $\varepsilon_{12}$ (all other strain components are zero). The only nonzero stress component is

$$\sigma_{12} = 2G\varepsilon_{12}. \tag{2.22}$$

The coefficient $G$ is called the *shear modulus*, or *modulus of rigidity*,

$$G = \mu = \frac{E}{2(1 + \nu)}. \tag{2.23}$$

*General uniform strain.* In the general case of isotropic material subjected to uniform deformation, the principal stresses and strains are related as follows:

$$\varepsilon_1 = \frac{1}{E}[\sigma_1 - \nu(\sigma_2 + \sigma_3)], \tag{2.24}$$

$$\sigma_1 = \frac{E}{(1 + \nu)(1 - 2\nu)}[(1 - \nu)\varepsilon_1 + \nu(\varepsilon_2 + \varepsilon_3)]. \tag{2.25}$$

## 2.10  THERMOELASTICITY

Consider a body in a reference (undeformed) state at temperature $T_0$. Increasing the temperature by a small amount $\Delta T = T - T_0$ in the absence of external body force usually results in expansion. The strain energy density $U$ must now be considered as a function of the strains and the temperature change $\Delta T$, which leads to the Taylor series expansion containing strain tensor components not only as quadratic but also as linear terms. For the isotropic body the only appropriate linear scalar combination is the sum $\Delta$ of normal strains $\varepsilon_{ij}$. For small $\Delta T$, the additional term in Eq. (2.16) may be assumed proportional to $\Delta T$, i.e.,

$$U = U_0 - K\beta\Delta T\varepsilon_{kk} + \frac{\lambda}{2}\varepsilon_{mm}^2 + \mu\varepsilon_{ij}^2, \tag{2.14'}$$

and

$$\sigma_{ij} = -K\beta\Delta T\delta_{ij} + 2\mu\varepsilon_{ij} + \lambda\delta_{ij}\varepsilon_{mm}. \tag{2.15'}$$

In a body expanding without constraint, no stresses arise, $\sigma_{ij} = 0$. It follows that $\varepsilon_{ij}$ must be proportional to $\delta_{ij}$, and, taking into account Eq. (2.15'), that $\varepsilon_{mm} = \delta_{ij}\beta\Delta T$, where $\beta$ represents the *volumetric thermal expansion coefficient*. Normal strain in any direction is given by $\varepsilon = \alpha\Delta T$, where $\alpha$ is the *linear thermal expansion coefficient*, and $\beta = 3\alpha$.

Anisotropy of thermal properties results in the orientation dependence of linear thermal expansion coefficients $\alpha_{ij}$ of crystals and polycrystals, whereby the thermal strains are given by

$$\varepsilon_{ij}^* = \alpha_{ij}\Delta T. \tag{2.26}$$

Here account is taken of the fact that, if global coordinate axes do not coincide with the principal axes of material symmetry, thermal expansion may cause apparent shear with respect to the global system.

The general thermoelastic constitutive law may be written in the form of modified Eq. (2.6) as

$$\sigma_{ij} = C_{ijkl}\left(\varepsilon_{kl} - \varepsilon_{kl}^*\right) = C_{ijkl}e_{kl}, \tag{2.6'}$$

where $e_{kl}$ denotes elastic strain; or in the form of modified Eq. (2.7) as

$$\varepsilon_{ij}^{\text{total}} = e_{ij} + \varepsilon_{ij}^* = s_{ijkl}\sigma_{kl} + \varepsilon_{ij}^*. \tag{2.7'}$$

Note that $e_{ij}$ has been used to refer to the elastic strain, alongside $\varepsilon_{ij}$ for the total strain, and $\varepsilon_{ij}^*$ for inelastic strain, or *eigenstrain*, represented in the particular case considered by the thermal strain. The first half of Eq. (2.7'),

$$\varepsilon_{ij} = e_{ij} + \varepsilon_{ij}^*, \tag{2.7''}$$

contains the expression of the additive relationship between these strain terms.

Eq. (2.26) shows that any inelastic strain $\varepsilon_{ij}^*(\mathbf{x})$ can be represented as a consequence of temperature change of, say, $\Delta T = 1$, provided the distribution of thermal expansion coefficients is chosen accordingly, i.e., $\alpha_{ij}(\mathbf{x}) = \varepsilon_{ij}^*(\mathbf{x})$. This observation provides an important basis for the use of *pseudothermal strain* to simulate *eigenstrain* distributions that serve as the underlying source of residual stress.

## 2.11 PLANE STRESS AND PLANE STRAIN

Plane problem of elasticity arises when deformation can be fully described in two-dimensional Cartesian coordinates. Two practically important cases are distinguished: plane stress and plane strain.

For *plane stress*, let $x_3$ be the direction normal to the surface of a thin flat plate loaded only along its edges. The out-of-plane stress components $\sigma_{13}$, $\sigma_{23}$, $\sigma_{33}$ can be neglected, and values of the in-plane stresses $\sigma_{11}$, $\sigma_{22}$, $\sigma_{12}$ be averaged through the thickness. Strains due to the in-plane stresses can be found from Eqs. (2.22) and (2.24):

$$\varepsilon_{12} = \frac{2(1+\nu)}{E}\sigma_{12}, \quad \varepsilon_1 = \frac{1}{E}[\sigma_1 - \nu\sigma_2], \quad \varepsilon_2 = \frac{1}{E}[\sigma_2 - \nu\sigma_1]. \tag{2.27}$$

The only nonzero out-of-plane strain is found to be $\varepsilon_3 = -\frac{\nu}{E}(\sigma_1 + \sigma_2)$.

*Plane strain* conditions arise if displacements everywhere in a solid body are perpendicular to the axis $Ox_3$ and do not depend on the corresponding coordinate $x_3$. Then from Eq. (2.17) strain components $\varepsilon_{13}$, $\varepsilon_{23}$, $\varepsilon_{33}$ vanish, and from Eqs. (2.22) and (2.24) it follows that $\sigma_{13} = \sigma_{23} = 0$ and $\sigma_{33} = \nu(\sigma_{11} + \sigma_{22})$. The strains are given by:

$$\varepsilon_{12} = \frac{2(1+\nu)}{E}\sigma_{12}, \quad \varepsilon_1 = \frac{(1-\nu^2)}{E}\left[\sigma_1 - \frac{\nu}{1-\nu}\sigma_2\right],$$

$$\varepsilon_2 = \frac{(1-\nu^2)}{E}\left[\sigma_2 - \frac{\nu}{1-\nu}\sigma_1\right]. \tag{2.28}$$

Note that Eq. (2.28) can be given the form equivalent to that of Eq. (2.27), if "plane strain elastic constants" are introduced:

$$E' = \frac{E}{(1-\nu^2)}, \quad \nu' = \frac{\nu}{1-\nu}. \tag{2.29}$$

This circumstance allows the use of the term *plane problem of elasticity* to refer both to plane stress and plane strain.

The Airy stress function provides a convenient means of expressing the solution of the plane problem of elasticity using one scalar biharmonic function $\phi(x, y)$: $\Delta\Delta\phi = 0$. It is defined by the formulas:

$$\sigma_{xx} = \frac{\partial^2\phi}{\partial y^2}, \quad \sigma_{yy} = \frac{\partial^2\phi}{\partial x^2}, \quad \sigma_{xy} = -\frac{\partial^2\phi}{\partial x\partial y}. \tag{2.30}$$

Adopting the Airy stress function ensures that both stress equilibrium and strain compatibility equations hold simultaneously, leaving only the functional form of $\phi(x, y)$ to be found, and boundary conditions to be satisfied. In the polar coordinates, the stress components are expressed in terms of the Airy stress function $\phi(r, \theta)$ as follows:

$$\sigma_{rr} = \frac{1}{r}\frac{\partial\phi}{\partial r} + \frac{1}{r^2}\frac{\partial^2\phi}{\partial\theta^2}, \quad \sigma_{\theta\theta} = \frac{\partial^2\phi}{\partial r^2}, \quad \sigma_{r\theta} = -\frac{\partial}{\partial r}\left(\frac{1}{r}\frac{\partial\phi}{\partial\theta}\right). \tag{2.31'}$$

## 2.12   FUNDAMENTAL RESIDUAL STRESS SOLUTIONS AND THE NUCLEI OF STRAIN

The mechanical behavior of real materials is strongly dependent on the presence within them of various defects, such as vacancies, dislocation lines and loops, and inclusions. These defects create self-equilibrated fields of stress and strain around them, which can be calculated on the basis of linear elasticity. The system of linear elastic equations introduced in the previous section admits fundamental singular solutions called *strain nuclei* (Mindlin, 1950), which play a role similar to the concentrated and distributed charges in electrostatics.

An example of a strain nucleus can be obtained readily by considering the Airy stress function in the form $\phi = C \log r$. By definition of stresses in terms of $\phi$ and the property of function being biharmonic, both equilibrium and compatibility conditions are satisfied. The stresses are given by

$$\sigma_{rr} = -\frac{C}{r^2}, \quad \sigma_{\theta\theta} = \frac{C}{r^2}. \tag{2.31}$$

Note that as $r \to \infty$, both stress components vanish, i.e., the solution corresponds to a residual stress state that persists in the absence of remote loading. Under plane strain conditions the strains are given by

$$\varepsilon_{rr} = -\frac{(1+\nu)}{E}\frac{C}{r^2}, \quad \varepsilon_{\theta\theta} = \frac{(1+\nu)}{E}\frac{C}{r^2}. \tag{2.32}$$

The only nonzero radial displacement component is $u_r = \frac{(1+\nu)}{E}\frac{C}{r}$. This strain nucleus solution is known as the *center of dilatation*.

Another example of a *strain nucleus* is given by an edge dislocation in an infinite solid. The boundary conditions are given in terms of the displacement discontinuity $u_2(x_2 + 0) - u_2(x_2 - 0) = b$, prescribed over the half plane $x_2 = 0$, $x_1 > 0$. The displacement jump $b$ is called the Burgers vector. The stress field around an edge dislocation is given by the following equations (Hull and Bacon, 1984), sometimes also referred to as the Sneddon equations:

$$\sigma_{11} = -\frac{Gb}{2\pi(1-\nu)}\frac{(3x_1^2 + x_2^2)x_2}{(x_1^2 + x_2^2)^2}, \quad \sigma_{22} = \frac{Gb}{2\pi(1-\nu)}\frac{(x_1^2 - x_2^2)x_2}{(x_1^2 + x_2^2)^2},$$

$$\tag{2.33}$$

$$\sigma_{33} = \nu(\sigma_{11} + \sigma_{22}), \quad \sigma_{12} = \frac{Gb}{2\pi(1-\nu)}\frac{(x_1^2 - x_2^2)x_1}{(x_1^2 + x_2^2)^2}, \quad \sigma_{23} = \sigma_{31} = 0.$$

The stresses are inversely proportional to the distance from the dislocation line in the plane $(x_1, x_2)$. The strain energy associated with the edge dislocation may be obtained by integration in the form $W = \ln(\Lambda/c)Gb^2/4\pi(1-\nu)$, where the outer dislocation radius $\Lambda$ and the dislocation core radius $c$ have to be introduced to avoid divergence. Although a continuum elastic description of defects such as dislocations breaks down in their immediate vicinity, it has been used successfully to model their interaction, as well as their effects on residual stress and deformation behavior. The subject of the elastic fields of dislocations, their interaction and motion, and their manifestation in the form of residual stresses is addressed in Chapter 8.

## 2.13 THE RELATIONSHIP BETWEEN RESIDUAL STRESSES AND EIGENSTRAINS

Residual stresses are the stresses that persist after the removal of the external loading and must therefore be self-equilibrating. Residual stresses can never arise because of elastic deformation alone but are caused by inelastic permanent strains, by additional constraints, by cutting and joining procedures, etc.

The principle of strain additivity expressed by Eq. (2.7″) allows the following generalization to be made. Using the term *eigenstrain* to refer to any strain $\varepsilon_{ij}^*$ inherited from inelastic deformation or processing history of the sample, we note that *eigenstrain theory* provides a complete and consistent basis for residual stress analysis. In a rational strain-based theory of inelastic deformation there are only two types of strains possible: elastic strains and eigenstrains. Only an inelastic process taking place in a solid (creep, phase transformation, plasticity, damage, or fracture) may modify the preexisting eigenstrain field. Theories of inelastic deformation, e.g., time-independent plasticity or creep relaxation, contain the decomposition of strain into elastic and inelastic strains (*eigenstrain*) as an essential step in the formulation. Another requisite aspect in the modeling of any inelastic process is prescribing the evolution law for eigenstrain in terms of the loading history (stress, time, temperature, etc.)

In a number of simple situations involving one-dimensional deformation (e.g., beam bending theory, axisymmetric deformation of thick-walled tubes under plane stress or plane strain), shell bending, etc., the *process modeling* approach is particularly effective, as the deformation problem can be solved analytically, and the residual elastic strains and residual stresses can be determined directly. These examples are considered in the next chapters.

For the purposes of general analysis of residual stresses a more efficient approach can be adopted, that later on in the presentation is referred to as *current state modeling*. The general *eigenstrain* theory of residual stresses rests on the assumption that the inelastic strain distribution is somehow known for the material state under analysis, from process modeling, destructive or nondestructive testing, or some other technique. The *direct problem of eigenstrain* analysis is posed as follows: for a given distribution of eigenstrain, find the elastic strains everywhere within the body, and hence calculate residual stresses.

It will be shown in the following chapters that this problem is "easy," in that it is linear, always has a unique solution, and can be solved without iteration. In fact, this problem corresponds to a perturbation of the strain compatibility equations that are part of the linear elastic problem formulation. Although in an elastic problem without eigenstrain these equations are homogeneous (have zero in the right-hand side), in the direct problem of eigenstrain the right-hand side is no longer zero but contains an expression that depends on eigenstrains.

In the cases of practical interest, however, the *inverse problem of eigenstrain* is encountered more frequently. This corresponds to the situation when some aspect of the residual stress state is known or measured, e.g., residual elastic strain values at a finite number of points can be determined from X-ray diffraction, or material relief (strain change) during layer removal or hole drilling that can be measured by strain gauges. The inverse problem requires the determination of an unknown eigenstrain distribution from the knowledge of some values of residual stresses, residual elastic strains, or their changes due to material removal. Once the eigenstrain distribution has been determined, the complete residual stress state can be reconstructed by a single solution of the direct problem.

The inverse problem of eigenstrain residual stress analysis is comparatively "hard," and may in fact be ill-posed. This means that its solution may not be unique and may change significantly even when the input data change by a small amount. The solution of such problems requires *regularization*, i.e., reformulation of the problem or restriction on the form of admissible trial functions, to ensure the existence and uniqueness of solution.

# CHAPTER 3

# Simple Residual Stress Systems

*Marcel Proust: Sur ce coteau normand etablis ta retraite, 1877*

*Alexander Frauchi (guitar): J.S. Bach, Chaconne from Partita no.2, BWV 1004.*

*J.M.W. Turner: High Street, Oxford (oil on canvas, 100 × 68 cm) 1810*
*Ashmolean Museum, Oxford*

## 3.1 ADDITIVITY OF TOTAL STRAIN

Eq. (2.7″) in the previous chapter gave the expression of the fundamental principle: the *total strain* is *additive*, i.e., is given by the sum of the *elastic* part $e_{ij}$ (that alone is responsible for and related to stresses) and the inelastic part $\varepsilon_{ij}^*$:

$$\varepsilon_{ij} = e_{ij} + \varepsilon_{ij}^*. \tag{2.7″}$$

In the derivation given in the previous chapter it was made clear that inelastic strain $\varepsilon_{ij}^*$ may correspond to thermal strain, but may in fact be of any origin, arising as a result of various deformation processes, such as creep, plasticity, and phase transformation. Constitutive law for material under consideration is needed to describe the evolution (often in incremental form) of the inelastic part of strain. This inelastic strain increment often depends on the applied stress that is directly linked to elastic strain through Hooke's law. Thus the principal function of any constitutive model is to partition the total strain increment (often prescribed) into the elastic and plastic parts. A classic example of constitutive law is the associated plastic flow rule.

The illustrative examples considered in this chapter relate to situations when an elastic–plastic body consists of *two* parts subjected to different loading or possessing different properties. With the stress state in each part being described by a single stress value, it is possible to obtain a number of situations when even with the overall external force being zero, nonzero internal stresses persist in each of the two parts.

**CONTENTS**

**21**

A Teaching Essay on Residual Stresses and Eigenstrains. http://dx.doi.org/10.1016/B978-0-12-810990-8.00003-3

## 3.2  CONSTRAINED ELASTIC–PLASTIC BAR LOADED AT A POINT ALONG ITS LENGTH

Consider an elastic-ideally plastic bar of length $L$, cross-sectional area $A$, Young's modulus $E$, and yield stress $\sigma_Y$ that is loaded by force $F$ at point $B$ along its length. The position of the loading point along the bar is defined by the parameter $\beta$, ($|AC| = \beta|AB| = \beta L$). For definiteness, assume that $\beta < 0.5$. The displacement of the point of load application is denoted by $d$. The stress–strain curve of the material is elastic-ideally plastic, as illustrated in Fig. 3.1.

As an aside, consider a change of temperature $\Delta T$ and assuming that the material's linear coefficient of thermal expansion is $\alpha$, we write the constraint equation in the form:

$$\varepsilon^{total} = e + \varepsilon^* = e + \alpha\Delta T = 0. \tag{3.1}$$

The internal stress in the bar that arises as a result of the combination of constraint and the introduction of thermally induced inelastic eigenstrain $\alpha\Delta T$ is given by

$$\sigma_R^1 = Ee = -E\alpha\Delta T. \tag{3.2}$$

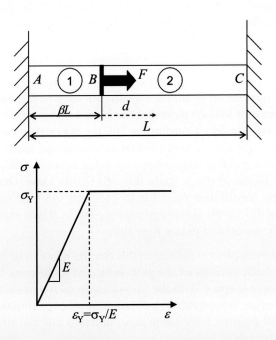

**FIGURE 3.1**

Constrained bar of unit cross-section loaded by force $F$ applied at fractional position $\beta$ along its length, and the elastic-ideally plastic material constitutive law.

We note that the situation is identical to one that would arise, were any other means used to introduce the inelastic strain (eigenstrain) $\varepsilon^*$, such as, e.g., cutting and pasting. Furthermore, positive (tensile) eigenstrain induces residual compression, and negative (compressive) eigenstrain causes residual tension.

We now consider an isothermal problem concerning the system response to an external loading by force $F$ at point $B$ that undergoes displacement $d$ (Fig. 3.1). Our particular interest is to consider the consequences of applying and then removing the external load, and evaluating the residual deformation that persists afterwards. First, we construct a simple model of the deformation process that allows the evaluation of the residual state. Later we discuss the possibility of using eigenstrain to obtain the same result.

When a displacement $d$ is applied at point $B$, different strains arise in the two parts of the bar designated regions 1 and 2, respectively, in Fig. 3.1. They are given by

$$\varepsilon^1 = \frac{d}{\beta L}, \quad \varepsilon^2 = -\frac{d}{(1-\beta)L}. \tag{3.3}$$

Note that $\varepsilon$ is used to denote *total strain*, and that compressive strain is caused in part 2 by positive displacement. The presence of the constraint ensures that the two strains must always satisfy the compatibility condition written as

$$\varepsilon^1 \beta L + \varepsilon^2 (1-\beta)L = 0, \tag{3.4}$$

so that the net displacement of point $C$ with respect to point $A$ remains zero.

Under the application of a gradually increasing force $F$, and as a function of the corresponding displacement $d$, three stages of deformation may be identified:

    Stage I: both parts of the bar remain elastic
    Stage II: part 1 becomes plastic, whereas part 2 remains elastic
    Stage III: both parts of the bar become fully plastic

For convenience, we introduce the shorthand notation $f = F/A$, and

$$s = Ed/L. \tag{3.5}$$

as a measure of displacement $d$ expressed in the units of stress. In terms of $s$, the strain expressions (3.1) are written as

$$\varepsilon^1 = \frac{s}{\beta E}, \quad \varepsilon^2 = -\frac{s}{(1-\beta)E}. \tag{3.6}$$

**Stage I:** $s < \beta\sigma_Y$, $f < \frac{\sigma_Y}{1-\beta}$.

Both parts of the bar remain elastic, so that the stresses in parts 1 and 2 are given by, respectively,

$$\sigma^1 = \frac{Ed}{\beta L} = \frac{s}{\beta}, \quad \sigma^2 = -\frac{Ed}{(1-\beta)L} = -\frac{s}{1-\beta}. \tag{3.7}$$

The total force within the bar in Stage I is given by the sum

$$f = \frac{F}{A} = \frac{F^1 - F^2}{A} = \sigma^1 - \sigma^2 = \frac{s}{\beta} + \frac{s}{1-\beta} = \frac{s}{\beta(1-\beta)}, \quad s = f\beta(1-\beta). \tag{3.8}$$

where $f$ denotes the force per unit area.

**Stage II**: $\beta\sigma_Y < s < \sigma_Y(1-\beta)$, $\frac{\sigma_Y}{1-\beta} < f < \frac{\sigma_Y}{\beta}$.

The onset of Stage II occurs when part 1 yields at $s = \beta\sigma_Y$, $f = \frac{\sigma_Y}{1-\beta}$. Stage II is completed when part 2 yields in compression at $s = \sigma_Y(1-\beta)$, $f = \frac{\sigma_Y}{\beta}$.

The strain expressions (3.3) derived earlier remain in force. Note, however, that the total strain $\varepsilon^1$ in part 1 is represented by the sum of the yield elastic strain of the material and *eigenstrain* $\varepsilon^*$ induced by plastic deformation:

$$\varepsilon^1 = e^1 + \varepsilon^* = \frac{\sigma_Y}{E} + \varepsilon^* = \frac{s}{\beta E}, \quad \varepsilon^* = \frac{s - \beta\sigma_Y}{\beta E}. \tag{3.9}$$

The stresses in part 1 and part 2 in Stage II are given by

$$\sigma^1 = \sigma_Y, \quad \sigma^2 = -s/(1-\beta). \tag{3.10}$$

The total force within the bar in Stage II is given by

$$f = \frac{F}{A} = \frac{F^1 - F^2}{A} = \sigma^1 - \sigma^2 = \sigma_Y + \frac{s}{1-\beta}, \quad s = (f - \sigma_Y)(1-\beta). \tag{3.11}$$

**Stage III**: $s > \sigma_Y(1-\beta)$, $f > \frac{\sigma_Y}{\beta}$.

Both part 1 and part 2 become fully plastic, so that $\sigma^1 = \sigma_Y$, $\sigma^2 = -\sigma_Y$, and the total force is

$$f = \frac{F}{A} = \frac{F^1 - F^2}{A} = \sigma^1 - \sigma^2 = 2\sigma_Y. \tag{3.12}$$

We are particularly interested in the residual stresses that persist in the bar after load removal. If the bar is unloaded from Stage I, no residual stresses arise, because the deformation of both parts was purely elastic. We focus our attention on the consideration of unloading from Stage II. We assume that no reverse plasticity takes place upon unloading, i.e., the deformation remains fully elastic. This assumption can be verified by checking that the stress states calculated for both parts of the bar during unloading conform to the requirement $|\sigma^i| < \sigma_Y$. For the purposes of calculation, we can assume that the unloading deformation increment is equal in magnitude but opposite in sign to the deformation that would occur in the bar under the same load, provided the bar remained elastic. Let the maximum force applied in Stage II be denoted by $\bar{F}$, so that

$$\bar{s} = (\bar{f} - \sigma_Y)(1 - \beta). \tag{3.13}$$

Considering unloading and using the elastic solution the displacement increment is found by inverting formula (3.8):

$$\Delta s = -\bar{f}\beta(1 - \beta). \tag{3.14}$$

The corresponding unloading stress increments in parts 1 and 2 of the bar are therefore, respectively,

$$\Delta\sigma^1 = -\bar{f}(1 - \beta), \quad \Delta\sigma^2 = \bar{f}\beta. \tag{3.15}$$

Combining (3.10) and (3.15) allows the calculation of the final residual stress state in the bar:

$$\sigma_R^1 = \sigma_Y - \bar{f}(1 - \beta) = \beta\sigma_Y - \bar{s}, \quad \sigma_R^2 = -\bar{f} + \sigma_Y + \bar{f}\beta = \beta\sigma_Y - \bar{s}. \tag{3.16}$$

As required, the residual stress state satisfies equilibrium:

$$f = \sigma_R^1 - \sigma_R^2 = 0. \tag{3.17}$$

Because in Stage II $\beta\sigma_Y < s$, compressive residual stress state arises following the introduction of inelastic permanent tensile plastic strain in part 1.

Note that the final state of the system can be obtained directly from the knowledge of the eigenstrain introduced by the process, as follows. Given the maximum load in Stage II described by parameters $\bar{s}$ and $\bar{f}$, the induced eigenstrain has the form given in Eq. (3.7):

$$\varepsilon^* = \frac{\bar{s} - \beta\sigma_Y}{\beta E}. \tag{3.18}$$

The residual stress state after unloading is characterized by $\sigma_R^1 = \sigma_R^2$, $e^1 = e^2 = e$. Hence, the compatibility condition (3.2) is written as

$$(\varepsilon^* + e)\beta + e(1 - \beta) = \frac{\bar{s} - \beta\sigma_Y}{E} + e = 0, \quad \text{and} \quad \sigma_R^1 = \sigma_R^2 = Ee = \bar{s} - \beta\sigma_Y. \tag{3.19}$$

The residual stress–strain state was evaluated using the knowledge of the underlying eigenstrain distribution, through direct use of the combination of equations of equilibrium and compatibility. This route to the solution is often shorter than the process modeling alternative that relies on following the evolution of the stress–strain state throughout the process. The eigenstrain approach becomes particularly relevant when the details of the process or of material constitutive behavior are not known.

The earlier calculations can be implemented readily in a simple spreadsheet, provided certain material properties are assumed. Letting the material of the bar to be similar to steel with Young's modulus of $E = 200$ GPa and yield stress of 1 GPa, let us assume the bar length to be 10 mm. For simplicity the bar cross-section can be taken to be 1 mm². Results of the calculation using these parameters are illustrated in Fig. 3.2.

Summarizing the results of this section we note that, even in the presence of external constraint, residual stresses require inelastic deformation to arise. In the presence of constraint, one such mode may be thermal expansion, or any other inelastic process that induces eigenstrain. The residual stress that arises

**FIGURE 3.2**

Illustration of the evolution (against displacement) of stresses in parts 1 and 2 of the bar, the total force $F$, and the residual stress that would arise after unloading. The bottom plot illustrates the response of the system subjected to cyclic loading and unloading.

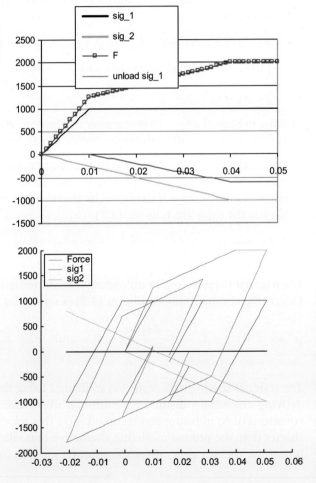

in this case after the removal of external load is *opposite in sign* to the inelastic strain introduced by the process (in part 1 of the bar in the present problem). We also note that the final residual stress—strain state can be evaluated using basic equations of continuum elasticity (stress equilibrium and strain compatibility), perturbed by the introduction of eigenstrain "forcing" terms.

**Question 3.1**: Prove that residual stresses in parts 1 and 2 must be equal.
**Question 3.2**: Compressive residual stresses arise in both parts of the bar. Specify loading sequence that would give rise to residual tension in the bar.
**Question 3.3**: Write a Matlab routine to calculate and plot bar deformation history. Comment on the form of the cyclic force-displacement relationships.

## 3.3  ELASTOPLASTIC COMPOSITES: UNIFORM STRESS (REUSS) AND STRAIN (VOIGT)

Consider a unidirectional layered composite consisting of two phases with different elastic and plastic properties: $E_1, \sigma_{Y1}$ and $E_2, \sigma_{Y2}$, respectively, and elastic-ideally plastic behavior illustrated in the lower plot in Fig. 3.1. The arrangement of phases is illustrated in Fig. 3.3, in which phase 1 is shown light, and phase 2 dark.

Let the volume fraction of phase 1 be denoted by $\beta$. The composite may be loaded externally either longitudinally or transversely. If the load is applied along the layers (longitudinally), then the deformation (strain) is the same

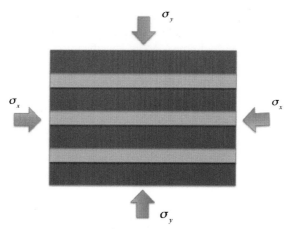

**FIGURE 3.3**
Multilayer composite subjected to transverse ($\sigma_y$) and longitudinal ($\sigma_x$) loading.

in both phases. This assumption is known as the uniform strain hypothesis, and associated with the name of Voigt. Alternatively, if the load is applied normally to the extension of layers (transversely), then the stress transmitted by the two phases may be assumed to be the same. The assumption of uniform stress in the system is associated with the name of Reuss.

Considering transverse loading by the stress $\sigma_y$ first, we note that the following relationship between stresses and strains arises:

$$\sigma_1 = E_1\varepsilon_1 = E_2\varepsilon_2 = \sigma_2. \tag{3.20}$$

Noting that

$$\varepsilon_C = \beta\varepsilon_1 + (1-\beta)\varepsilon_2 = \left[\frac{\beta}{E_1} + \frac{1-\beta}{E_2}\right]\sigma = \frac{\sigma}{E_C}, \tag{3.21}$$

the composite Young's modulus is obtained in the form

$$E_C = 1 \Big/ \left[\frac{\beta}{E_1} + \frac{1-\beta}{E_2}\right]. \tag{3.22}$$

This relationship referred to as the inverse rule of mixtures is often also used as a bound for stiffness estimation in cases when the geometry of the system is more complex than that shown in Fig. 3.3, e.g., for transverse modulus of unidirectional fiber-reinforced composites.

Our interest in this case concerns the onset of inelastic behavior, and subsequent unloading. Yielding will occur when the applied stress equals the smaller of the phase yield stresses: $\sigma_{YC} = \min(\sigma_{Y1}, \sigma_{Y2})$. If the materials are elastic-ideally plastic, plastic flow will follow with no further increase in stress. After unloading, residual stresses in both phases will be zero, despite the presence of plasticity-induced eigenstrain in one of the phases. This important observation highlights the fact that the presence of eigenstrain does not automatically result in residual stress. More specifically, the situation occurs when residual deformation has a configuration in which the total strain satisfies compatibility without requiring elastic strain for accommodation. As a consequence, elastic strains remain zero everywhere, and no residual stresses arise.

We now consider loading in the longitudinal direction, i.e., stress $\sigma_x$. The deformation that arises in this case corresponds to the uniform strain (Voigt) approximation:

$$\varepsilon_1 = \varepsilon_2 = \varepsilon = \frac{\sigma_1}{E_1} = \frac{\sigma_2}{E_2}. \tag{3.23}$$

**Stage I**: $\varepsilon < \varepsilon_{Y1}$, $\sigma_C < \frac{E_C}{E_1}\sigma_{Y1}$.

At Stage I the deformation remains elastic, and the composite stress and stiffness are found by the rule of mixtures:

$$\sigma_C = \beta\sigma_1 + (1 - \beta)\sigma_2 = [\beta E_1 + (1 - \beta)E_2]\varepsilon = E_C\varepsilon. \tag{3.24}$$

**Stage II**: $\varepsilon_{Y1} < \varepsilon < \varepsilon_{Y2}$, $E_C\varepsilon_{Y1} < \sigma_C < \beta\sigma_{Y1} + (1 - \beta)\sigma_{Y2}$.

The yield strains for the two phases are given by $\varepsilon_{Y1} = \sigma_{Y1}/E_1$ and $\varepsilon_{Y2} = \sigma_{Y2}/E_2$. For definiteness let us assume that $\varepsilon_{Y1} < \varepsilon_{Y2}$, so that phase 1 yields first. The composite stress is given by

$$\sigma_C = \beta\sigma_{Y1} + (1 - \beta)E_2\varepsilon. \tag{3.25}$$

Owing to the attainment of maximum stress $\sigma_{Y1}$ in phase 1, the slope of the composite stress–strain curve drops from the value of $E_C = \beta E_1 + (1 - \beta)E_2$ in Stage I to $(1 - \beta)E_2$ in Stage II.

**Stage III**: $\varepsilon > \varepsilon_{Y2}$, $\sigma_C > \beta\sigma_{Y1} + (1 - \beta)\sigma_{Y2}$.

Ideally plastic behavior is observed in this stage, with composite stress saturating at the value $\sigma_{Cmax} = \beta\sigma_{Y1} + (1 - \beta)\sigma_{Y2}$ that is intermediate between phase yield stresses $\sigma_{Y1}$ and $\sigma_{Y2}$.

The stress–strain plots for the two-phase Voigt composite are illustrated in Fig. 3.4, where the responses of individual phases and the composite medium are indicated. One interesting aspect is the emergence of strain hardening behavior: the composite stress–strain curve displays a gradual transition from linear elastic behavior to stress saturation.

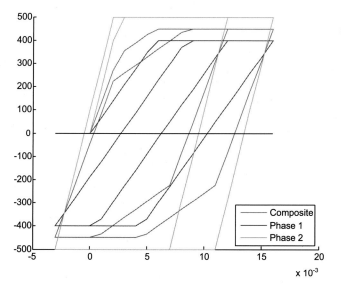

**FIGURE 3.4**

Composite and phase stresses in the Voigt approximation.

Another interesting aspect is the emergence of kinematic hardening that manifests itself in the so-called Bauschinger effect, i.e., the reduction of the material (composite) yield stress in compression following yielding in tension. The explanation for this effect can be readily provided in terms of internal interphase residual stresses, or, equivalently, eigenstrains.

Consider unloading from a load point in Stage II that is characterized by the parameters $\bar{\varepsilon}$ and $\bar{\sigma}_C = \beta\sigma_{Y1} + E_2\bar{\varepsilon}$. Phase 1 strain can be decomposed into the elastic and inelastic (eigenstrain) parts:

$$\bar{\varepsilon} = e + \varepsilon^* = \varepsilon_{Y1} + (\bar{\varepsilon} - \varepsilon_{Y1}). \tag{3.26}$$

We now consider elastic unloading by strain $-\Delta\varepsilon$ that happens in accordance with the rules of Stage I, i.e., so that

$$\Delta\sigma_1 = -E_1\Delta\varepsilon, \quad \Delta\sigma_2 = -E_2\Delta\varepsilon. \tag{3.27}$$

and overall

$$\sigma_1 = \sigma_{Y1} - \Delta\sigma_1 = \sigma_{Y1} - E_1\Delta\varepsilon, \quad \sigma_2 = \bar{\sigma}_2 - \Delta\sigma_2 = E_2(\bar{\varepsilon} - \Delta\varepsilon). \tag{3.28}$$

The composite stress is given by

$$\sigma_C = \beta\sigma_1 + (1-\beta)\sigma_2 = \beta\sigma_{Y1} + (1-\beta)E_2\bar{\varepsilon} - E_C\Delta\varepsilon. \tag{3.29}$$

Residual stress state is found by setting the composite stress to zero, giving

$$E_C\Delta\varepsilon = \beta\sigma_{Y1} + (1-\beta)E_2\bar{\varepsilon}, \tag{3.30}$$

so that finally, the residual stresses are given by

$$\tilde{\sigma}_1 = (1-\beta)\frac{E_2}{E_C}(\sigma_{Y1} - E_1\bar{\varepsilon}), \quad \tilde{\sigma}_2 = \frac{E_2}{E_C}\beta\sigma_{Y1} + E_2\bar{\varepsilon}\left(1 - (1-\beta)\frac{E_2}{E_C}\right)$$

$$= \beta\frac{E_2}{E_C}(\sigma_{Y1} + E_1\bar{\varepsilon}). \tag{3.31}$$

It is apparent that residual stresses in the two phases are of opposite signs, with phase 1 that has undergone yielding in tension finding itself in residual compression (since in Stage II $E_1\bar{\varepsilon} > \sigma_{Y1}$).

This observation of compressive bias in the softer phase 1 provides the explanation of the Bauschinger effect. As unloading proceeds into compression, yielding occurs when

$$\Delta\sigma_1 = -2\sigma_{Y1}, \quad \Delta\varepsilon = \frac{2\sigma_{Y1}}{E_1}. \tag{3.32}$$

Therefore, the change in the composite stress required for the reverse yielding to occur is given by:

$$\Delta\sigma_C = -E_C\Delta\varepsilon = -\frac{2E_C}{E_1}\sigma_{Y1}. \tag{3.33}$$

After simplification, the composite stress for reverse yielding is found to be

$$\sigma_C = \bar{\sigma}_C - \Delta\sigma_C = -\frac{E_C}{E_1}\sigma_{Y1} + (1-\beta)\frac{E_2}{E_1}(E_1\bar{\varepsilon} - \sigma_{Y1}). \tag{3.34}$$

In the last expression, the first term is the opposite of the composite yield stress in tension, whereas the second term is strictly positive by virtue of maximum strain in tension $\bar{\varepsilon}$ lying in Stage II. The Bauschinger effect can therefore be directly linked to the generation and influence of internal residual stresses.

## 3.4 ON THE COMPOSITE MECHANICS OF POLYCRYSTALS

By weight and volume, most of the structural materials in use are polygranular. This means that they are composed of distinct domains of consistent material properties ("*grains*"), in terms of composition, structure, orientation, etc., that are joined together across *grain boundaries*. Taking the example of concrete, one can identify distinct phases (e.g., cement and aggregate) and pose questions about the average values of strain, stress, plastic deformation, damage, and residual stress in each phase. Developing this idea further we note that different properties can be used to define phases. The key differentiating property in the example considered in the previous section was the yield strain $\varepsilon_Y$. Because two distinct values of this parameter were present in the system, $\varepsilon_{Y1}$ and $\varepsilon_{Y2}$, two phases were present. The structure shown in Fig. 3.3 can be thought of as a representative volume element within which the phases are present in volume fractions of $\beta$ and $(1-\beta)$, respectively.

In real polycrystalline alloys, the distribution of yield strains is caused by the dependence of the flow stress of crystallites on the orientation of the underlying lattice structure with respect to the loading direction (plastic anisotropy), as well as the anisotropy of elastic stiffness, and the effects of constraint imposed by the grain structure surrounding a particular crystal. Some aspects of the deformation conditions experienced by grains within a polycrystal are reflected in the examples considered earlier, namely, constraint offered to plastically deforming grains by the surrounding elastic "hinterland," as well as aspects of stress transmission (Reuss approximation, emphasizing stress equilibrium) and strain sharing (Voigt approximation, emphasizing continuity). Polycrystalline deformation must conform to the continuum mechanics laws, so that both stress equilibrium and strain compatibility are satisfied within any material volume. Capturing these complex interactions within a simple model is not possible, although considerable progress has been made using methods based on so-called *self-consistent methods* and upper and lower bound solutions.

From the earlier discussion it follows that the onset of yield within polycrystals occurs over a range of strains, serving as one of the causes of macroscopically manifest continuous strain hardening. In the illustration provided in Fig. 3.5A, the two-step "ladder" corresponds to the two-phase Voigt composite considered in the previous section. The piecewise constant function describes the cumulative distribution function (CDF) of the material yielded under tensile strain $\varepsilon$ (abscissa). The case of a 5-phase composite is illustrated by a piecewise constant "many step ladder" function indicated by the dashed line. A continuous distribution of yield strain within the material can be represented by a monotonically nondecreasing continuous function $\beta(\varepsilon)$ (solid curve). This function represents the volume fraction of the material yielded upon the attainment of strain $\varepsilon$.

Let us derive the relationship between $\beta(\varepsilon)$ and the composite stress–strain curve. For simplicity let us assume that all phases have the same Young's modulus, so that the stress–strain curve $\sigma(e)$ for the phase with yield strain $\varepsilon$ is given by

$$\sigma(e) = \begin{cases} Ee, & e < \varepsilon \\ E\varepsilon, & e > \varepsilon \end{cases}. \tag{3.35}$$

Referring to Fig. 3.5B, consider the range of strains between $e$ and $e + de$. The volume fraction of material that undergoes yielding in this range is $\beta'(e)de$. Now let us compute the composite stress when the strain reaches $\varepsilon$. If no yielding took place during prior deformation, the stress would be given by the elastic relation, $\sigma = E\varepsilon$. Yielding causes a reduction in the load-bearing capacity of the volume fraction that became plastic. In particular, the

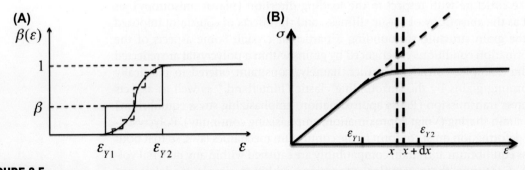

**FIGURE 3.5**
(A) Cumulative distribution function of the yield volume fraction $\beta(\varepsilon)$ versus strain. (B) Illustration for the derivation of the composite stress–strain curve.

contribution to this reduction caused by the *additional* yielding of volume fraction $\beta'(e)\mathrm{d}e$ between $e$ and $e + \mathrm{d}e$ is equal to $E(\varepsilon - e)\beta'(e)\mathrm{d}e$. Therefore, the overall composite stress at $\varepsilon$ is found by integration:

$$\sigma(\varepsilon) = E\varepsilon - E \int_0^\varepsilon (\varepsilon - e)\beta'(e)\mathrm{d}e. \tag{3.36}$$

Integrating by parts yields the following expression:

$$\sigma(\varepsilon) = E\varepsilon(1 + \beta(0)) - E \int_0^\varepsilon \beta(e)\mathrm{d}e. \tag{3.37}$$

Differentiating both sides with respect to $\varepsilon$ gives the following relation:

$$\beta(\varepsilon) = 1 - \frac{1}{E}\frac{\mathrm{d}\sigma}{\mathrm{d}\varepsilon}. \tag{3.38}$$

This expression allows exploring the relationship between the CDF of the yield volume fraction of the material, $\beta(\varepsilon)$, on the one hand, and the stress–strain curve $\sigma(\varepsilon)$, on the other. As an example, consider the following definition:

$$\beta(\varepsilon) = \arctan\left(A\left(\frac{\varepsilon}{\varepsilon_0} - 1\right)\right)\Big/\pi + \frac{1}{2}. \tag{3.39}$$

This function is illustrated in Fig. 3.6A for $A = 5\pi$.

Fig. 3.6B shows the corresponding normalized stress–strain curve.

The simple composite model presented here has limited utility but enjoys the advantages of being tractable and transparent. Furthermore, it can be used to explore the evolution of internal residual stress under monotonic and cyclic loading. This treatment may be seen as related to the Masing hypothesis (Masing, 1923), which states that the reverse yield curve (and the cyclic hysteresis loop) of a metal can be obtained from the monotonic tensile yield curve by reversing the signs and doubling the magnitude of the increments of strain and stress with respect to the reversal point.

> Question 3.4: Evaluate the phase residual stresses following inelastic stretching of the continuous yield composite to maximum strain $\bar{\varepsilon}$.
> Question 3.5: Examine the relationship between the continuous yield composite model and the Masing hypothesis.

## 3.5  THE RAMBERG–OSGOOD STRESS–STRAIN RELATIONSHIP

The Ramberg–Osgood form of the stress–strain relationship is based on three fundamental hypotheses:

**FIGURE 3.6**

Illustration of the relationship between (A) the cumulative distribution function of the yield volume fraction versus normalized yield strain $\varepsilon/\varepsilon_0$, and (B) the material stress–strain curve, also plotted in the normalized form, as $\sigma/E\varepsilon_0$ versus $\varepsilon/\varepsilon_0$.

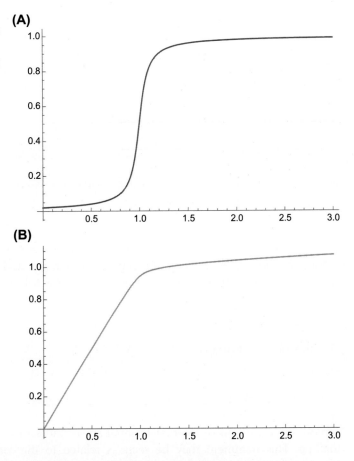

1. Elastic strain is proportional to applied stress: $\varepsilon_{el} = \frac{\sigma}{E}$.
2. Plastic strain is given by a power law function of applied stress:
$$\varepsilon_{pl} = \left(\frac{\sigma}{K}\right)^n.$$
3. Total strain is given by the sum of elastic and plastic strains:

$$\varepsilon = \varepsilon_{el} + \varepsilon_{pl} = \frac{\sigma}{E} + \left(\frac{\sigma}{K}\right)^n. \tag{3.40}$$

Experimental stress–strain data can be readily approximated by the Ramberg–Osgood form of stress–strain relationship. First, the slope of the linear elastic part is determined and allows the determination of Young's modulus $E$. Second, the plastic strain is computed as $\varepsilon_{pl} = \varepsilon - \varepsilon_{el} = \varepsilon - \frac{\sigma}{E}$. Finally, the unknown parameters $K$ and $n$ are determined by fitting a power law relationship $\varepsilon_{pl} = \left(\frac{\sigma}{K}\right)^n$ to the experimental data. For example, taking logarithms of both sides we find $\ln\varepsilon_{pl} = n\ln\sigma - n\ln K$. Hence

$$n = \frac{\ln\left(\varepsilon_{pl,2}/\varepsilon_{pl,1}\right)}{\ln(\sigma_2/\sigma_1)} \quad \text{and} \quad K = \frac{\sigma}{\varepsilon_{pl}^{1/n}}. \tag{3.41}$$

One of the notable properties of the Ramberg-Osgood relationship is the absence of sharply defined yield. Therefore, nominally plastic strain is present under the application of any stress, no matter how small. However, in practice the choice of the power exponent $n$ allows control over hardening behavior. The yield stress may be estimated, e.g., via a construction similar to 0.2% proof stress: $\varepsilon_{pl} = 0.002 \approx \left(\frac{\sigma_Y}{K}\right)^n$, $\sigma_Y \approx K(0.002)^{1/n}$.

In practical applications it is often convenient to use the incremental form of the Ramberg–Osgood relationship. This is obtained as follows:

$$\frac{d\varepsilon}{d\sigma} = \frac{1}{E} + \left(\frac{1}{K}\right)^n n\sigma^{n-1} = \frac{1}{E} + \left(\frac{\sigma}{K}\right)^n \frac{n}{\sigma}, \quad \text{and} \quad \text{hence} \quad \frac{d\sigma}{d\varepsilon} = \frac{1}{\left(\frac{1}{E} + \left(\frac{\sigma}{K}\right)^n \frac{n}{\sigma}\right)}. $$
$$\tag{3.42}$$

Provided the increment of strain used in the description is small enough, numerical integration of the incremental form of the Ramberg–Osgood relationship provides a sufficiently accurate description of the stress–strain curve. This is illustrated in Fig. 3.7.

The greatest value of the Ramberg–Osgood form of the stress–strain relationship lies in the fact that it provides an accurate and flexible description of the curves for a wide variety of materials and loading conditions, including

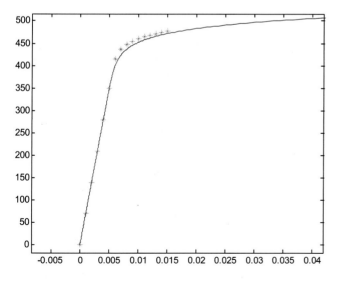

**FIGURE 3.7**
Illustration of the Ramberg–Osgood stress–strain relationship (*continuous curve*) and the stepwise integration of its incremental form (*markers*).

monotonic and cyclic loading. The generalization of this description to three-dimensional cases with the possibility of nonproportional and reverse loading requires further analysis.

Ramberg–Osgood material deformation law may also be considered in the context of cyclic loading. Note, however, that this requires redefining the starting point for this stress–strain curve as the point of load reversal, sic:

$$\varepsilon - \varepsilon_{ref} = \frac{\sigma - \sigma_{ref}}{E} + \text{sgn}(\sigma - \sigma_{ref})\left(\frac{|\sigma - \sigma_{ref}|}{K_{cyc}}\right)^{n}. \tag{3.43}$$

On the basis of the *Masing hypothesis* mentioned in the previous section, the stress–strain curve doubles in magnitude upon load reversal. This hypothesis was explained by considering the continuum as an aggregate (composite) of groups of elastic-ideally plastic phases (grain groups) that possess a range of yield stresses. In the context of Ramberg–Osgood description this implies that upon load reversal $K_{cyc} = 2K$. This type of response is illustrated in Fig. 3.8. In practice it is observed for many structural metallic alloys that both parameters $K$ and $n$ need to be replaced with cyclic values $K_{cyc}, n$. This allows capturing of both the kinematic and isotropic forms of hardening.

The relationship between the Ramberg–Osgood description and the previously introduced continuously yielding composite model is illustrated in Fig. 3.9. For the purpose of calculation, the following values of parameters were taken: $E = 1$, $K = 1$, $n = 10$. The form of the CDF of yield volume fraction reveals the progression of yielding implied by this model, and confirms that the Ramberg–Osgood stress–strain relationship can be understood in terms of the continuous composite model, and internal residual stress generation and influence on subsequent deformation.

**FIGURE 3.8**
Illustration of the Ramberg–Osgood cyclic stress–strain relationship.

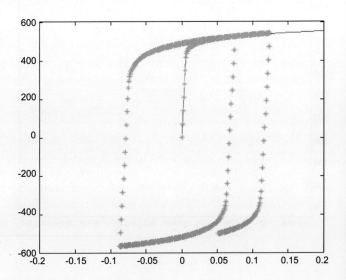

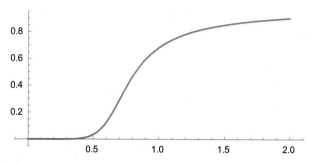

**FIGURE 3.9**

Illustration of the cumulative distribution function yield volume fraction as a function of total strain, computed on the basis of the Ramberg–Osgood stress–strain relationship (see text).

## 3.6 CONTINUUM PLASTICITY

Let us return to the continuum description of deformation: the material we are considering may possess internal structure, e.g., may be a continuously yielding composite, but for the purposes of analysis in the present section we assume that initially the material is homogeneous and uniform everywhere. When a cylindrical specimen is subjected to tensile stretching beyond its elastic limit, it undergoes plastic deformation. As indicated in Fig. 3.10, the onset yielding can be described in terms of either yield stress $\sigma_Y$ or yield strain $\varepsilon_Y$. As we have seen previously, *strain hardening* is a term used to describe further increase of stress during deformation beyond the elastic limit, and if the stress remains constant in this range, the material was said to be *elastic-ideally plastic*.

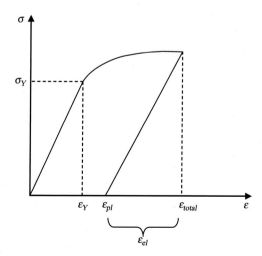

**FIGURE 3.10**

Inelastic stretching of a tensile specimen.

Sometimes *softening* behavior may be observed, associated with stress reduction during plastic flow.[1]

Indicated in Fig. 3.10 are also the total strain $\varepsilon_{total}$ and its partitioning into the elastic $\varepsilon_{el}$ and plastic $\varepsilon_{pl}$ parts. Sometimes $\varepsilon_{pl}$ is also called inelastic strain, permanent strain, or *eigenstrain*, and is assumed to persist when elastic unloading begins.

The situation depicted in Fig. 3.10 corresponds to the simplest loading case when both the material and the stress distribution within it are assumed to be homogeneous. Therefore, the state of stress is characterized by a single parameter, $\sigma$, that must be equal to applied stress $\sigma_{appl} = F/A$, where $F$ is the externally applied load. When external load is removed, then $\sigma = \sigma_{appl} = 0$. Therefore, no residual stress may be present in the absence of inhomogeneity of material properties or stress distribution (e.g., due to sample geometry). The study of residual stresses requires the analysis of spatially nonuniform stress and strain distributions, and/or of inhomogeneity of material properties. We have seen in the previous section that the generation of internal residual stress during deformation depends on the presence of a material *structure*, i.e., a certain arrangement of mechanically, chemically, and physically distinguishable phases.

Tensile tests of the type shown in Fig. 3.10 provide information about the conditions under which initial onset of yield takes place, and also about the material behavior beyond yield (evolution of stress and strain, elastic–plastic partitioning).

Continuum plasticity theory is concerned with the description of stresses, and elastic and inelastic strains in two-dimensional and three-dimensional elastic–plastic bodies. It is evident from the discussion in the previous chapter that stresses and strains (both elastic and inelastic) must be described by tensor quantities. Therefore, the criterion for the onset of plastic deformation must be generalized, as must also be the constitutive equations that describe the evolution of strains.

The von Mises yield *criterion* is formulated in terms of the components of stress tensor as follows:

$$\overline{\sigma} \leq \sigma_Y, \quad \text{where} \quad \overline{\sigma} = \sigma_{vM}(\boldsymbol{\sigma}) = \sqrt{\left(3\sigma_{ij}\sigma_{ij} - \sigma_{kk}\right)}/\sqrt{2}. \tag{3.44}$$

where $\overline{\sigma}$ is the equivalent von Mises stress. It may also be written in terms of the principal stresses as follows:

---

[1] Note to avoid confusion the same terminology may also be applied to describe *cyclic behavior*.

$$\bar{\sigma} = \sqrt{1/2}\sqrt{(\sigma_{11}-\sigma_{22})^2 + (\sigma_{22}-\sigma_{33})^2 + (\sigma_{11}-\sigma_{33})^2}. \tag{3.45}$$

Note that similarly the equivalent plastic strain can be defined as follows:

$$\bar{\varepsilon} = \sqrt{2/3}\sqrt{\varepsilon_{ij}^{pl}\varepsilon_{ij}^{pl}}. \tag{3.46}$$

The coefficients $\sqrt{1/2}$ and $\sqrt{2/3}$ are set by calibration against uniaxial yielding.

The three-dimensional yield criterion is formulated by assuming that, for any stress state, the onset of yielding occurs when $\bar{\sigma}$ reaches $\sigma_Y$. Equation

$$\gamma_p = \bar{\sigma}(\boldsymbol{\sigma} - \boldsymbol{\sigma}_0, \boldsymbol{\varepsilon}_{pl}) - \sigma_Y(\boldsymbol{\varepsilon}_{pl}) = 0 \tag{3.47}$$

defines the so-called *yield surface*. This is a surface in the principal stress space that corresponds to the locus of points where yielding takes place (Fig. 3.11).

The evolution of stresses and strains in the course of plastic deformation is governed by the so-called *flow rule* that consists of the following postulates:

1. At the onset of yielding the increments of the stress and strain tensors are collinear.
2. The plastic strain increment is normal to the yield surface:

$$d\boldsymbol{\varepsilon}_p = d\lambda \frac{d\gamma}{d\boldsymbol{\sigma}}. \tag{3.48}$$

For von Mises plastic yield criterion, the flow rule can be expressed as follows:

$$d\boldsymbol{\varepsilon}_p = \frac{3}{2}d\lambda \frac{\boldsymbol{\sigma}^D}{\bar{\sigma}}. \tag{3.49}$$

here $\boldsymbol{\sigma}^D$ is the deviatoric part of the stress tensor given by $\boldsymbol{\sigma}^D = \boldsymbol{\sigma} - \frac{1}{3}(\mathrm{tr}\boldsymbol{\sigma})\mathbf{I}$, where $\mathrm{tr}\boldsymbol{\sigma} = \sigma_{11} + \sigma_{22} + \sigma_{33}$, and $\mathbf{I}$ is the diagonal unit tensor. In the analysis

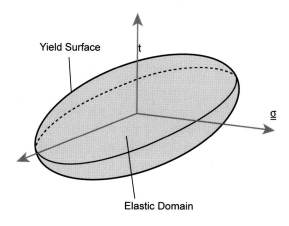

**FIGURE 3.11**
Illustration of the von Mises yield surface.

of plastic deformation a distinction is made between plastic loading and elastic unloading. In other words, plastic flow occurs *only* if the yield criterion is satisfied *and* the stress increment points outward with respect to the yield surface (if not, elastic unloading takes place).

The *consistency condition* is used to determine the magnitude of the *plastic multiplier* d$\lambda$. This condition reflects the fundamental hypothesis that the points in the stress space that represent the material states before and after a deformation increment that involves plastic flow must find themselves not outside, but on the yield surface. Hence

$$\gamma_p\left(\boldsymbol{\sigma}, \boldsymbol{\varepsilon}^{pl}\right) = \gamma_p\left(\boldsymbol{\sigma} + d\boldsymbol{\sigma}, \boldsymbol{\varepsilon}^{pl} + d\boldsymbol{\varepsilon}^{pl}\right) = 0. \tag{3.50}$$

Expanding the right-hand side into series and simplifying, the above equation gives

$$\frac{d\gamma_p}{d\boldsymbol{\sigma}} d\boldsymbol{\sigma} + \frac{d\gamma_p}{d\boldsymbol{\varepsilon}^{pl}} d\boldsymbol{\varepsilon}^{pl} = 0, \quad \text{and} \quad d\boldsymbol{\varepsilon}^{pl} = d\lambda\sqrt{\frac{2}{3}\frac{d\gamma_p}{d\boldsymbol{\sigma}}\frac{d\gamma_p}{d\boldsymbol{\sigma}}}. \tag{3.51}$$

This equation allows the determination of plastic strain increment. The knowledge of the stress increment that preserves the consistency condition (3.50) allows the elastoplastic tangent stiffness matrix to be determined in the form:

$$\mathbf{C}_{el-pl} = \left(\mathbf{C} - \mathbf{C}\cdot\frac{d\gamma_p}{d\boldsymbol{\sigma}}\ \frac{\mathbf{C}\cdot\dfrac{d\gamma_p}{d\boldsymbol{\sigma}}}{\dfrac{d\gamma_p}{d\boldsymbol{\sigma}}\cdot\left(\mathbf{C}\cdot\dfrac{d\gamma_p}{d\boldsymbol{\sigma}} - \sqrt{\dfrac{2}{3}\dfrac{d\gamma_p}{d\boldsymbol{\sigma}}\dfrac{d\gamma_p}{d\boldsymbol{\sigma}}}\right)}\right). \tag{3.52}$$

Note that in Eq. (3.47) both the von Mises equivalent stress and the yield stress are written as functions of the plastic strain tensor, $\boldsymbol{\varepsilon}_{pl}$. This assumption reflects the observation that real materials undergo complex hardening (here used as a generic term referring to the evolution of yield stress with deformation) that results in the modification of the yield surface. Conventionally the types of hardening behavior considered include isotropic and kinematic hardening. Isotropic hardening corresponds to the yield surface expansion in all directions that results in the modification of the yield stress $\sigma_Y(\boldsymbol{\varepsilon}_{pl})$. Kinematic hardening causes yield surface translation as it is "dragged" by the loading point through the stress space. It leads to the presence of the so-called *back stress* $\boldsymbol{\alpha}$; $\boldsymbol{\alpha}$ and $\sigma_Y(\boldsymbol{\varepsilon}_{pl})$ are *internal variables* that evolve in the course of deformation. Their evolution is often prescribed in the incremental form, as

$$d\sigma_Y = S(\boldsymbol{\sigma}, \boldsymbol{\varepsilon}_{pl}, d\boldsymbol{\varepsilon}_{pl}), \quad d\boldsymbol{\alpha} = K(\boldsymbol{\sigma}, \boldsymbol{\varepsilon}_{pl}, d\boldsymbol{\varepsilon}_{pl}). \tag{3.53}$$

# CHAPTER 4

# Inelastic Bending of Beams

*Rainer Maria Rilke: Der Lesende 1901*

*Robert Schumann, Piano concerto no.1 in A minor, 1845 (end of 1st movement)*

*Edgar Degas, Blue Dancers (pastel, 65 × 65 cm) 1899 Pushkin Museum, Moscow*

## 4.1 SLENDER RODS: COLUMNS, BEAMS, AND SHAFTS. SAINT-VENANT'S PRINCIPLE

Beams and columns, also referred to as slender rods (Fig. 4.1), are a geometric idealization of a deformable solid that applies when one of the dimensions of the body (length) is significantly larger than the other two (thickness $d$ and width $b$). Slenderness of the rod implies that $\max(b, d) \ll L$.

When an externally applied load acts along the length of the rod (in this case it is called a *column*), then a uniaxial state of stress arises: the only nonzero component of stress is the *axial* stress $\sigma_{xx}$; all other components are zero. Moreover, far away from the region of application of external load the stress distribution across any section of the slender rod by a plane $x = $ const is uniform, i.e., does not depend on the coordinates $y$ and $z$, provided the rod is a constant cross-section prism, i.e., can be obtained by extruding the section along the $x$ direction. However, close to the region of the load application, for $x \leq x_0 \approx \max(b, d)$, the stress distribution is dependent on the details of the distribution of applied tractions. The term "close" here refers to the characteristic dimension of the cross-section, taken to be the larger of $b$ and $d$.

This statement is the simplest manifestation of the so-called Saint-Venant's principle. Saint-Venant's principle also applies to other slender members: when the rod is loaded by end moments $M_y$ or $M_z$, or transverse forces $F_y$ or $F_z$, it is called a *beam*; when the rod is loaded by an end torque $M_y$ it is called a *shaft*.

When a column experiences axial stress that exceeds its yield stress $\sigma_Y$, it undergoes plastic deformation and acquires macroscopic permanent plastic strain (eigenstrain) $\varepsilon^*$ that is also uniform; $\varepsilon^* = $ const. When the external loading is

**41**

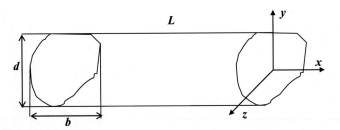

**FIGURE 4.1**
A slender rod.

removed, the eigenstrain persists. Notably, no residual elastic strain or residual stress arises in this case. We note in this example that the introduction of *eigenstrain* is a necessary but not sufficient condition for the creation of residual stress. *Inhomogeneous* eigenstrain is needed to generate residual stress. If the material of the column is itself inhomogeneous and the resulting eigenstrain distribution is nonuniform at the microscopic scale, then *microscopic residual stresses* arise.

Perhaps the simplest case of inhomogeneous eigenstrain that creates macroscopic residual stress is that arising in a *beam* that undergoes inelastic bending, discussed in the next section.

## 4.2  INELASTIC BEAM BENDING

A basic formulation of the elastic beam bending theory is introduced in Fig. 4.2. In this formulation, uniform bending moment $M = M_z$ is applied, and only axial displacement $u(x)$ and axial elastic strain $e_{xx}$ are thought to take place. The only stress component that develops is the axial stress $\sigma_{xx}$. Under uniaxial elastic deformation conditions the relationship between stress and strain is given by

$$\sigma_{xx} = Ee_{xx}, \tag{4.1}$$

where $E$ is Young's modulus. A brief summary of the basic principles of elastic beam bending is summarized at the bottom of Fig. 4.2 in the form of mnemonic known as the *"Most Important Statement You Ever Remember"*.

Under elastic bending conditions both strain and stress vary linearly with coordinate $y$. Yielding first occurs when the greatest stress at the beam surface reaches the yield stress $\sigma_Y$. The analysis of inelastic bending of a rectangular beam made from an elastic-ideally plastic material is illustrated in Fig. 4.3.

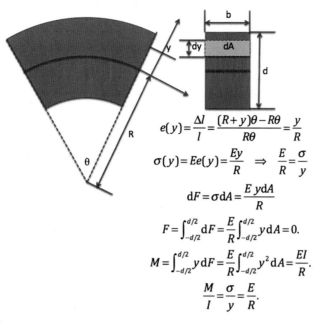

$$e(y) = \frac{\Delta l}{l} = \frac{(R+y)\theta - R\theta}{R\theta} = \frac{y}{R}$$

$$\sigma(y) = Ee(y) = \frac{Ey}{R} \quad \Rightarrow \quad \frac{E}{R} = \frac{\sigma}{y}$$

$$dF = \sigma dA = \frac{E\,y dA}{R}$$

$$F = \int_{-d/2}^{d/2} dF = \frac{E}{R}\int_{-d/2}^{d/2} y\,dA = 0.$$

$$M = \int_{-d/2}^{d/2} y\,dF = \frac{E}{R}\int_{-d/2}^{d/2} y^2\,dA = \frac{EI}{R}.$$

$$\frac{M}{I} = \frac{\sigma}{y} = \frac{E}{R}.$$

**FIGURE 4.2**

Elastic beam bending.

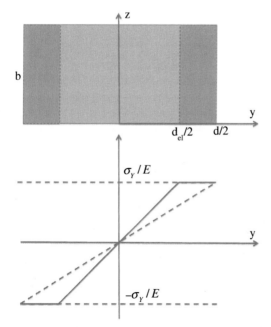

**FIGURE 4.3**

Inelastic beam bending.

First yielding occurs when the maximum strain at the surface of the beam reaches the yield strain $\varepsilon_Y = \sigma_Y/E$, and the stress is $\sigma = \frac{Ed/2}{R} = \sigma_Y$. The yielding moment of the beam is therefore given by $M_Y = \frac{2\sigma_Y I}{d}$. The corresponding curvature is given by $\kappa_Y = \frac{1}{R_Y} = \frac{2\sigma_Y}{Ed}$, and defines the limit of the linear relationship between moment and curvature. When a beam is bent more tightly than that, the moment varies nonlinearly. We will show that it approaches the asymptotic limit given by $M_{\max} = 1.5M_Y$ for elastic-ideally plastic beam of rectangular cross-section (Timoshenko and Goodier, 1965).

The behavior of a beam subjected to bending beyond the elastic limit can be described in terms of the relationship between the neutral axis curvature (which is equal to the total strain gradient within the beam) and the applied moment. Alternatively, one can use the parameter that describes the relative position of the elastic–plastic boundary, $r = d_{\text{el}}/d$. Yielding under uniaxial tensile or compressive stress first occurs at the periphery of the beam cross-section (Fig. 4.3). For the stage of inelastic bending depicted, the relationship between curvature and the extent of elastic core is given by

$$\kappa = \frac{2\sigma_Y}{Ed_{\text{el}}} = \frac{M}{EI}.$$

The calculation of the total applied moment $M$ proceeds as follows. For $|y| < d_{\text{el}}/2$, the elastic core of the beam, the bending moment carried can be found using the elastic bending formulae, leading to $M_{\text{el}} = \frac{\sigma_Y b d_{\text{el}}^2}{6}$. For the plastic zones $|y| < d_{\text{el}}/2$, stress is everywhere equal in magnitude to $\sigma_Y$, and the corresponding "plastic" contribution to the bending moment is given by

$$M_{\text{el}} = 2 \times \sigma_Y b \frac{(d-d_{\text{el}})}{2} \times \frac{(d+d_{\text{el}})}{4} = \sigma_Y b \frac{(d^2-d_{\text{el}}^2)}{4}.$$ The resultant moment is given by

the sum as $M = M_{\text{el}} + M_{\text{pl}} = 1.5M_Y\left(1 - \frac{d_{\text{el}}^2}{3d^2}\right) = 1.5M_Y\left(1 - \frac{r^2}{3}\right)$.

Plastic deformation causes the introduction of permanent plastic strain (*eigenstrain*) within the plastic zones lying at the beam surfaces on the tensile and compressive sides. Because *eigenstrain* distributions are nonuniform, after unloading residual stress arises.

Two procedures can be employed to determine the residual stress induced by inelastic bending. In the traditional formulation, elastoplastic analysis of bending is first carried out and then elastic unloading is superimposed (e.g., Timoshenko and Goodier, 1965). The other approach favored here is to consider the residual stress to arise in response to the eigenstrain generated during loading, and to develop a procedure for the determination of residual elastic strain and residual stress in response to this eigenstrain distribution. This example of the *direct problem of residual stress analysis* is discussed in the next section.

**Question 4.1:** Evaluate and plot the relationship between applied moment and beam curvature for an elastic-ideally plastic beam of rectangular cross-section. What is the beam curvature required for complete yielding of a beam in bending?

**Question 4.2:** Considering elastic unloading, evaluate the amount of beam "springback" (in terms of curvature) following plastic bending.

**Question 4.3:** Describe the procedure for evaluating the profile of residual stress that persists in a plastically bent beam after unloading. Explain clearly the assumptions made.

## 4.3 DIRECT PROBLEM: RESIDUAL STRESS IN A PLASTICALLY BENT BEAM

Consider an elastic beam occupying the region $x_L < x < x_R$, $-\infty < y < \infty$ and containing a distribution of eigenstrain that varies within the cross-section, but not along the beam length, $\varepsilon^*_{xx} = \varepsilon^*(y)$.

The following statements provide the basis for the analysis:

1. Total strain in the beam is given by the sum of the elastic and inelastic strain (eigenstrain).
2. Following Kirchhoff's hypothesis of straight normals, it is assumed that material points originally lying on a line perpendicular to the beam axis remain on a straight line, i.e., any normal to the beam axis only undergoes rotation without distortion.
3. Hence displacements, and therefore *total* strain, must vary linearly through the plate thickness, i.e., is given by

$$\varepsilon = e + \varepsilon^* = a + bx/h, \tag{4.2}$$

   where $h = x_R - x_L$ is the beam thickness. Here the parameter $a$ characterizes the amount of axial straining experienced by the beam, and term $b$ characterizes the intensity of bending.
4. In the absence of external loading being applied, elastic strain $e$ presents an example of macroscopic residual elastic strain, such as that measured in a diffraction experiment.
5. From Eq. (4.2), residual elastic strain is given by

$$e = a + bx/h - \varepsilon^*(x), \tag{4.3}$$

   If the dependence of parameters $a$ and $b$ on the eigenstrain distribution $\varepsilon^*(x)$ is known, then the relationship between the residual elastic strain $e$ and the eigenstrain $\varepsilon^*(x)$ is established.
6. It will be shown (later) that parameters $a$ and $b$ depend solely on two integral parameters, namely, the zeroth and first moments of the eigenstrain distribution given by

$$\Gamma = \frac{1}{h}\int_{x_L}^{x_R} \varepsilon^*(x)\mathrm{d}x, \quad \Gamma_1 = \frac{1}{h^2}\int_{x_L}^{x_R} \varepsilon^*(x)x\mathrm{d}x \tag{4.4}$$

The relationship between parameters $a$ and $b$, on the one hand, and $\Gamma$ and $\Gamma_1$ on the other is established using the requirements of force and moment balance across the beam, given by

$$F = \int_{x_L}^{x_R} [a + bx/h - \varepsilon^*(x)]\mathrm{d}x = 0, \tag{4.5}$$

$$M = \int_{x_L}^{x_R} [a + bx/h - \varepsilon^*(x)]x\mathrm{d}x = 0. \tag{4.6}$$

leading to the following relationships:

$$(x_R + x_L)b/2 + a(x_R - x_L) - (x_R - x_L)\Gamma = 0, \tag{4.7}$$

$$\left(x_R^2 + x_R x_L + x_L^2\right)b/3 + \left(x_R^2 - x_L^2\right)a/2 - (x_R - x_L)^2\Gamma_1 = 0. \tag{4.8}$$

Expressions are given explicitly in terms of the beam boundaries $x_L$ and $x_R$ for the purposes of generality, e.g., to allow the consideration of effects of surface layer removal.

The solution of the linear system for parameters $a$ and $b$ has the form

$$a = \frac{6\Gamma_1\left(x_R^2 - x_L^2\right) - 4\Gamma\left(x_R^2 + x_R x_L + x_L^2\right)}{(x_R - x_L)^2}, \tag{4.9}$$

$$b = \frac{12\Gamma_1(x_R - x_L) - 6\Gamma(x_R + x_L)}{(x_R - x_L)}. \tag{4.10}$$

Noting that because bending component of strain in terms of beam bending radius $R$ and the beam curvature $\kappa$ is given by

$$e = \frac{x}{R} = x\kappa, \tag{4.11}$$

then from Eq. (4.3) the curvature of the bent beam is found as

$$\kappa = \frac{b}{h} = \frac{12\Gamma_1(x_R - x_L) - 6\Gamma(x_R + x_L)}{(x_R - x_L)^2}, \tag{4.12}$$

Eq. (4.12) contains an expression that is useful for the analysis of beam curvature as a function of the eigenstrain distribution $\varepsilon^*(x)$.

Substituting Eqs. (4.9) and (4.10) back into Eq. (4.2) gives the resulting prediction for the residual elastic strain distribution in the form

$$e(x) = \frac{1}{(x_R - x_L)^2}\left[6\Gamma_1(x_R - x_L)(2x - x_R - x_L) + 2\Gamma\left((x_R^2 + x_R x_L + x_L^2\right)\right.$$
$$\left. - 3x(x_R + x_L))\right] - \varepsilon^*(x). \tag{4.13}$$

Eq. (4.13) establishes the solution of the direct problem about the determination of residual elastic strain for arbitrary given distribution of eigenstrain.

Fig. 4.4 gives an example of the eigenstrain distribution arising as a result of inelastic bending of a beam made from elastic-ideally plastic material. The horizontal coordinate is normalized with respect to the beam half-thickness $d/2$, and the vertical coordinate is normalized with respect to the yield strain $\varepsilon_Y = \sigma_Y/E$. Fig. 4.5 illustrates the profile of the residual elastic strain, $e$. Owing to the simple linear relationship between elastic strain and axial stress this profile also describes the distribution of the normalized residual stress.

In practice, few materials display true elastic-ideally plastic behavior, so that inelastic bending of beams made from such materials induces more complex residual stress states than the piecewise linear profile shown in Fig. 4.5. A problem that often arises in practice concerns the determination of the underlying eigenstrain state that gives rise to the residual elastic strain that is somehow measured experimentally. This problem is addressed in Chapter 9.

## 4.4 CASE: RESIDUAL STRESSES DUE TO SURFACE TREATMENT

An interesting application of the approach presented in the previous section arises in the context of surface treatment, when a mechanical, thermal, or chemical modification of the near-surface regions is induced to modify the material properties (e.g., nitriding, carburizing, induction hardening, or coating) or to create residual compression near the surface that is beneficial to the object's fatigue resistance.

Laser shock peening is a surface treatment technique that relies on inducing a shock compression wave propagating from the material surface by rapid discharge of energy within an absorptive layer of, say, carbon. The passage of the shock wave causes plastic deformation and induces a shallow distribution

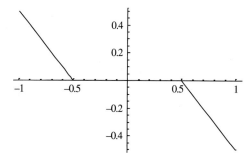

**FIGURE 4.4**

Eigenstrain distribution induced by inelastic bending of an elastic-ideally plastic beam.

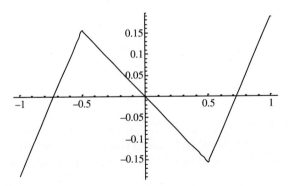

**FIGURE 4.5**
Elastic strain distribution in a residually bent beam.

of eigenstrain (to about 2−3 mm below surface) that in turn creates a state of residual compression. Because the resulting residual stress state must obey equations of equilibrium, the corresponding residual elastic strain distribution penetrates deeper into the material and must contain both compressive and tensile regions. This situation is similar to that arising from other, more conventional treatment methods, such as shot peening and low plasticity burnishing.

The results of the previous section provide a convenient framework for the analysis of induced residual stress. It is convenient in many situations to characterize the process directly in terms of eigenstrain distribution that it introduces. To illustrate this principle we postulate a distribution of eigenstrain, and compute the residual elastic strain and residual stress states that arise in response.

We begin analysis by considering the process of "slicing" off a beam element from a laser shock peened plate. The system of coordinates in Fig. 4.6 is consistent with the definitions of the previous section. Note that this operation leads to the relaxation of some stress components, e.g., $\sigma_{zz}$. In the present analysis we ignore this effect, because within bending theory no account is taken of this stress component, or of the interaction between mutually orthogonal strain components via the Poisson effect. We note, however, that a treatment similar to that presented for *beams* can be developed for *plates* and *shells*, i.e., elements possessing a thickness dimension much smaller than the other two.

We introduce an eigenstrain distribution $\varepsilon^*{}_{xx} = \varepsilon^*(y)$ that is representative of laser shock peening. A suitable function must reach a maximum value some way below the surface and must fall smoothly to zero at certain depth. The calculation of the residual elastic strain and stress following the formulae of the previous section can be readily implemented in a spreadsheet. Fig. 4.7 illustrates an example.

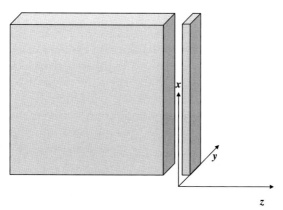

**FIGURE 4.6**
Extraction of a beam element by "slicing" a laser shock peened plate.

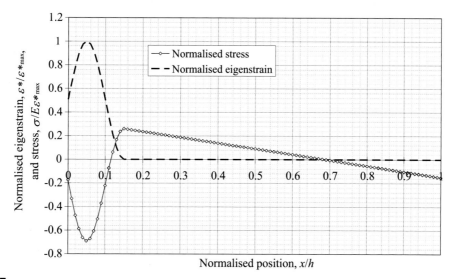

**FIGURE 4.7**
Example of normalized eigenstrain and normalized residual stress in a beam extracted from a laser shock peened plate. The horizontal coordinate has been normalized with respect to the plate thickness.

## 4.5 CASE: RESIDUAL STRESSES IN COATINGS AND THIN LAYERS

The treatment described previously finds an important application in the analysis of residual stresses in coatings and thin layers. A simple model based on direct eigenstrain calculation can be readily constructed by assuming a stratified material system consisting of $i = 1..N$ layers, each occupying the domain $x_{i-1} < x < x_i$ and possessing Young's modulus value $E_i$ and thermal expansion coefficient $\alpha_i$. The model allows straightforward estimation of residual stresses that arise because of the mismatch in the thermal expansion coefficient between different coating layers and the substrate. Note that growth stresses that arise within the system during coating deposition may also be treated in this way, by making additional assumptions about the eigenstrain evolution with deposition history. To illustrate the approach, the presentation below draws on the paper devoted to the study of residual stresses in microcantilever sensors (Korsunsky et al., 2007).

The total strain within the system $\varepsilon$ is given by the sum of the elastic and inelastic (eigenstrain) constituents, $\varepsilon = e + \varepsilon^*$. The Kirchhof hypothesis of straight normal, which enforces a particular form of strain compatibility for this system, requires that the total strain vary linearly across the cross-section of the coated cantilever, i.e.,

$$\varepsilon = a + by. \tag{4.14}$$

Term $a$ here refers to the component of strain uniform across the section, whereas the coefficient $b$ describes the strain gradient (slope) and is related to the radius of curvature of the cantilever, $R$, by the simple expression $b = 1/R = y''$. The residual stress within each layer $i$ is simply related to the elastic part of total strain,

$$\sigma_i = E_i e_i = E_i(\varepsilon - \varepsilon_i^*) = E_i(a + by - \varepsilon_i^*). \tag{4.15}$$

This provides the basis for developing a simple thermomechanical deformation model and the determination of parameters $a$ and $b$, and hence the prediction of cantilever curvature and its change with temperature.

The eigenstrain model is constructed as follows. The conditions of equilibrium for the multicoated system are expressed in terms of the resultant force and moment per unit width of the cantilever. Assuming a trilayer system as an example, we can write:

$$F = \int_{x_0}^{x_3} \sigma(y) dy = \int_{x_0}^{x_1} E_1(a + by) dy + \int_{x_1}^{x_2} E_2(a + by - \varepsilon_2^*) dy$$
$$+ \int_{x_0}^{x_1} E_3(a + by - \varepsilon_3^*) dy = 0. \tag{4.16}$$

$$M = \int_{x_0}^{x_3} \sigma(y) y dy = \int_{x_0}^{x_1} E_1(a + by) y dy + \int_{x_1}^{x_2} E_2(a + by - \varepsilon_2^*) y dy$$
$$+ \int_{x_0}^{x_1} E_3(a + by - \varepsilon_3^*) y dy = 0. \tag{4.17}$$

The solution for parameter $b$ (and hence the radius of curvature) can be written in the form

$$b = \frac{2B_N}{B_D},$$

where

$$
\begin{aligned}
B_N = {} & [E_1(x_1 - x_0) + E_2(x_2 - x_1) + E_3(x_3 - x_2)] \times [E_2\varepsilon_2^*(x_2^2 - x_1^2) \\
& + E_3\varepsilon_3^*(x_3^2 - x_2^2)] - [E_1(x_1^2 - x_0^2) + E_2(x_2^2 - x_1^2) + E_3(x_3^2 - x_2^2)] \\
& \times [E_2\varepsilon_2^*(x_2 - x_1) + E_3\varepsilon_3^*(x_3 - x_2)]
\end{aligned}
\tag{4.18}
$$

$$
\begin{aligned}
B_D = {} & E_1(x_1^2 - x_0^2) + E_2(x_2^2 - x_1^2) + E_3(x_3^2 - x_2^2)^2 + (4/3)[E_1(x_1 - x_0) \\
& + E_2(x_2 - x_1) + E_3(x_3 - x_2)] \times [E_1(x_1^3 - x_0^3) + E_2(x_2^3 - x_1^3) \\
& + E_3(x_3^3 - x_2^3)]
\end{aligned}
\tag{4.19}
$$

From the dependence of $b$ on eigenstrains $\varepsilon_2^*$ and $\varepsilon_3^*$, and above-mentioned equation, it is apparent that curvature $b$ scales linearly with eigenstrain. Therefore, adding to the above treatment an evolution law for eigenstrain(s) as a function of layer deposition/removal or time, system curvature and residual stress can be predicted and compared with observation.

> **Question 4.4:** Evaluate the parameter $a$ in the form of Eqs. (4.18) and (4.19), and demonstrate its linear dependence on eigenstrain.
> **Question 4.5:** Investigate the dependence of residual stress in a coated system on the substrate thickness, and represent the result graphically.
> **Question 4.5:** Referring to the predictions of eigenstrain model, discuss the function of a thin bond layer often introduced between a thicker coating and substrate.

# CHAPTER 5

# Plastic Yielding of Cylinders

*Louisa Stuart Costello: Oh, who can listen with delight to tales of battles won? 1814*

*Franz Schubert, Sonata in A major D664, 1825*

*Karl Bryullov, The Last Day of Pompeii (oil on canvas, 651 × 466 cm)*
*1833, Russian Museum, St Petersburg*

## 5.1 INELASTIC EXPANSION OF A THICK-WALLED TUBE

The inelastic expansion of a thick-walled tube under internal pressure is a prominent example of a residual stress generation mechanism that is associated with *eigenstrain* arising as a consequence of plastic flow.

The solution for *elastic* axisymmetric deformation of a thick-walled tube of internal radius $a$ and external radius $b$ under internal pressure $p$ is a special case of the Lamé problem (Fig. 5.1).

The following formulae give the radial stress $\sigma_r$ and hoop stress $\sigma_\theta$ (Den Hartog, 1949):

$$\sigma_r^0(r) = -p\frac{(b^2 - r^2)a^2}{(b^2 - a^2)r^2}, \quad \sigma_\theta^0(r) = p\frac{(b^2 + r^2)a^2}{(b^2 - a^2)r^2}. \tag{5.1}$$

Now assume that the internal pressure $p$ applied is sufficient to cause plastic yielding in the inner annulus of the tube, $a < r < c$, while the outer part of the tube, $c < r < b$, remains elastic. The solution can then be split into two parts by assuming that the radial traction transmitted across the cylindrical boundary $r = c$ is given by the as yet unknown pressure $q$ (assumed to be a positive number for convenience). Owing to the problem's symmetry the shear tractions vanish everywhere, $\sigma_{r\theta} = \sigma_{rz} = 0$, and the following boundary conditions apply:

Inner plastic region:

$$\sigma_r(a) = -p, \quad \sigma_r(c) = -q, \tag{5.2}$$

A Teaching Essay on Residual Stresses and Eigenstrains. http://dx.doi.org/10.1016/B978-0-12-810990-8.00005-7

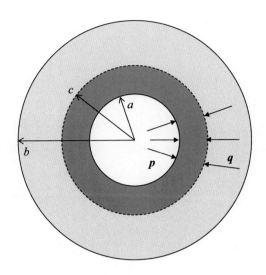

**FIGURE 5.1**
Geometry of a thick-walled tube under internal pressure.

Outer elastic region:

$$\sigma_r(c) = -q, \quad \sigma_r(b) = 0. \tag{5.3}$$

The solution for the outer elastic region can then be readily found by modification of Eq. (5.1) using the following substitutions:

$$p \to q, \quad a \to c, \tag{5.4}$$

resulting in

$$\sigma_r^e(r) = -q\frac{(b^2 - r^2)c^2}{(b^2 - c^2)r^2}, \quad \sigma_\theta^e(r) = q\frac{(b^2 + r^2)c^2}{(b^2 - c^2)r^2}, \quad c \le r \le b. \tag{5.5}$$

The solution for the plastic region is developed as follows. Using Tresca's criterion it is assumed that everywhere within the plastic region the following yield condition is satisfied:

$$\sigma_\theta - \sigma_r = \sigma_Y, \tag{5.6}$$

where $\sigma_Y$ is the yield stress in uniaxial tension. Note that this implies the assumption that the axial stress component $\sigma_z$ takes an intermediate value between $\sigma_\theta$ and $\sigma_r$. This assumption can be verified once the solution is obtained.

The above-mentioned equation can be solved together with the radial equilibrium equation:

$$\frac{\partial \sigma_r}{\partial r} + \frac{\sigma_r - \sigma_\theta}{r} = 0. \tag{5.7}$$

Satisfying the first of the boundary conditions (Eq. 5.2) gives the following expressions for stresses in the inner plastic region (Den Hartog, 1949):

$$\sigma_r^p = \sigma_Y \log\frac{r}{a} - p, \quad \sigma_\theta^p = \sigma_Y\left(\log\frac{r}{a} + 1\right) - p, \quad a \le r \le c. \tag{5.8}$$

Using the second of the boundary conditions (Eq. 5.2) an expression for $q$ is obtained:

$$q = p - \sigma_Y \log\frac{c}{a}. \tag{5.9}$$

Finally, noting that the yield condition (Eq. 5.6) must also be satisfied at the boundary $r = c$ of the outer elastic region, the implicit equation for the extent of the plastic zone, $c$, is obtained in terms of the problem parameters:

$$p = \frac{\sigma_Y}{2}\left(1 - \frac{c^2}{b^2} + 2\log\frac{c}{a}\right). \tag{5.10}$$

Eqs. (5.5) and (5.8) together provide the complete stress solution for the problem of elastoplastic deformation of a thick-walled tube under internal pressure.

To obtain the expressions for residual stresses and residual elastic strains, we now consider the process of elastic unloading. The stresses given by Eqs. (5.5) and (5.8) are modified by subtracting the solution for purely elastic stresses given in Eq. (5.1), so that the final residual stress state has the form

$$\sigma_r^R = \sigma_r^p - \sigma_r^0, \quad \sigma_\theta^R = \sigma_\theta^p - \sigma_\theta^0, \quad a \le r \le c, \tag{5.11}$$

$$\sigma_r^R = \sigma_r^e - \sigma_r^0, \quad \sigma_\theta^R = \sigma_\theta^e - \sigma_\theta^0, \quad c \le r \le b. \tag{5.12}$$

Assuming plane stress conditions, the residual elastic strains can be written in the form

$$Ee_r(r,c,p,a,b,\sigma_Y) = \sigma_r^R - v\sigma_\theta^R, \quad Ee_\theta(r,c,p,a,b,\sigma_Y) = \sigma_\theta^R - v\sigma_r^R. \tag{5.13}$$

Figs. 5.2–5.4 present examples of the distributions of stresses and strains arising from the model. The particular values of model parameters chosen here are $a = 1, b = 3, \sigma_Y = 100$ MPa. The plots are presented for the purposes of illustration.

Fig. 5.2 shows the hoop and radial stress distribution with radial position at maximum loading. The hoop stress reaches a peak value at $r = 70$ mm, which in this illustrative example corresponds to the position of the elastic–plastic boundary, $c$.

Fig. 5.3 illustrates the residual stresses after unloading (indicated by the overbar).

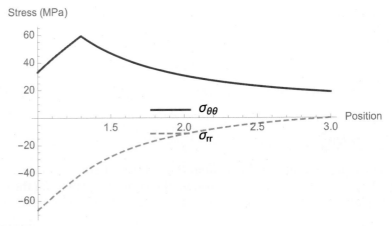

**FIGURE 5.2**
Elastic-ideally plastic model prediction of stress under internal pressure sufficient to cause plastic deformation.

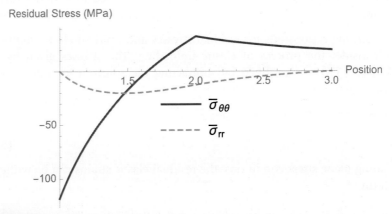

**FIGURE 5.3**
Example elastic-ideally plastic model predictions for stress distributions after unloading.

Fig. 5.4 shows the residual elastic strains calculated using Young's modulus of 70 GPa and Poisson's ratio of 0.3.

Formulae given in this section provide the complete solution of the direct problem about the determination of residual elastic strains from given geometry $(a, b)$, material properties ($\sigma_Y$, Young's modulus $E$, and Poisson's ratio $\nu$), and loading conditions (pressure $p$). In other words, the residual elastic strain profiles $\varepsilon_{rr}$ and $\varepsilon_{\theta\theta}$, in a rather general sense, can be thought of as functions of the problem parameters $c, p, a, b, \sigma_Y$. This view provides a basis for developing an approach to solving the inverse problem of autofrettage.

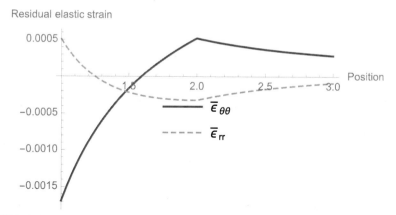

**FIGURE 5.4**
Example elastic-ideally plastic model predictions for residual elastic strains.

## 5.2 CASE: AUTOFRETTAGED TUBES AND COLD-EXPANDED HOLES

The interpretation of experimental measurements of residual elastic strains (or their changes) may be conducted for identifying the underlying deformation parameters that control the residual stress state that arises as a consequence. In the case of an axially symmetric inelastic deformation of thick-walled tubes subjected to an internal pressure that exceeds the elastic limit of the tube (autofrettage), the a priori unknown parameter is the radial position of the elastic–plastic boundary, $c$. Furthermore, if deformation takes place at an elevated temperature, the material yield stress may also not be known. The problem that arises in this case may be formulated as follows:

> Let the radial and hoop components of the residual elastic strain be known as a function of radial position in an autofrettaged tube from measurements, e.g., by diffraction. Determine the extent of plastic yield (the radial position of the elastic–plastic boundary, $c$) and the material yield stress by matching the model to the experiment.

To seek the best agreement between the prediction and the measurements, it is necessary to define a measure of misfit and express it as a function of input parameters, $J(c, \sigma_Y)$. A convenient definition of this function can be cast in the form of the sum of squares:

$$J(c, \sigma_Y) = \sum_{j=1}^{N} \left[ w_{rr}^j \left( \varepsilon_{rr}^j - e_{rr}^j \right)^2 + w_{\theta\theta}^j \left( \varepsilon_{\theta\theta}^j - e_{\theta\theta}^j \right)^2 \right].$$

Here $\varepsilon_{rr}^j$, $\varepsilon_{\theta\theta}^j$ denote the experimentally measured strain radial and hoop strain components, respectively, and $e_{rr}^j$, $e_{\theta\theta}^j$ denote the model predictions (implicit functions of parameters $c$ and $\sigma_Y$). The choice of weights $w_{rr}^j$ and $w_{\theta\theta}^j$ can be made on the basis of additional information available, e.g., as the inverse of the standard deviation of individual measurements of the radial and hoop strain components, respectively. The solution of the problem formulated earlier was presented in (Korsunsky, 2007a) by direct evaluation and minimization of the misfit function $J(c, \sigma_Y)$. The contour plot of this function is shown in Fig. 5.5. The corresponding match between the model prediction and experimental residual elastic strain data is shown in Fig. 5.6.

Cold expansion is routinely applied to the holes used for riveted joints in the aerospace industry to improve the resistance of assemblies to crack initiation and propagation. The mechanism of eigenstrain and residual stress generation is well described by the simple model introduced earlier. A particular version of the cold expansion treatment uses a "dimpling" treatment that involves compression of a plate between two spherical indenters that causes the increase of pressure and outward plastic flow of material. The process leads to the introduction of axially symmetric eigenstrain distribution, and after the removal of the central core by drilling, the remaining residual elastic strain is similar to that arising following autofrettage and cold expansion treatment.

Fig. 5.7 illustrates a match between the optimized model and the residual elastic strain measured experimentally using synchrotron X-ray diffraction (Zhang et al., 2008a).

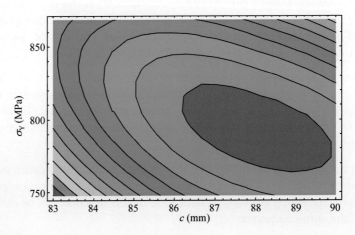

**FIGURE 5.5**

Contour plot of the sum-of-squares two-parameter misfit function $J(c, \sigma_Y)$ between the model prediction and the measured residual elastic strains (Korsunsky, 2007a).

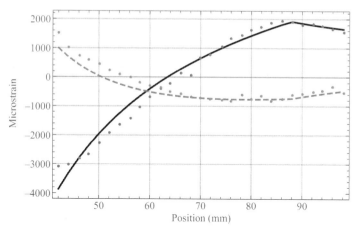

**FIGURE 5.6**

Illustration of the match between the model prediction and the measured residual elastic strains using a two-parameter model (Korsunsky, 2007a).

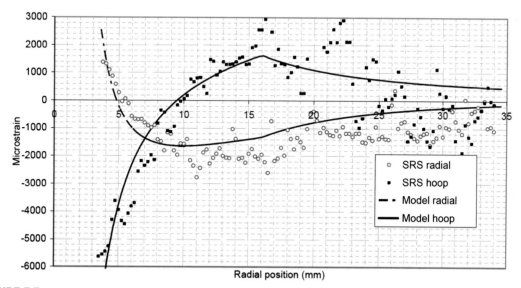

**FIGURE 5.7**

Illustration of the match between strain prediction of the model and experimental measurements by synchrotron X-ray diffraction conducted at Synchrotron Radiation Source (SRS), Daresbury. The significant scatter is associated with large grain size of the sample material. (Zhang et al., 2008a).

## 5.3   CASE: QUENCHING OF A SOLID CYLINDER

Consider a cylinder of radius $a$ of uniform initial temperature $T_1$ that is subjected to surface cooling (Korsunsky et al., 2016). We assume that the surface temperature is reduced to $T_0$ at time $t = 0$, and maintained at that level afterward. The temperature distribution within the cylinder is governed by the transient heat conduction equation:

$$\frac{\partial T}{\partial t} = \alpha \frac{1}{r} \frac{\partial}{\partial r} \left( r \frac{\partial T}{\partial r} \right) \tag{5.14}$$

that has the solution expressed in terms of Bessel's functions as follows:

$$\frac{T(r,t) - T_0}{T_1 - T_0} = 2 \sum_{n=1}^{\infty} \frac{J_0(\lambda_n \rho)}{\lambda_n J_1(\lambda_n)} \exp(-\lambda_n^2 \tau), \tag{5.15}$$

where the normalized radial coordinate and normalized time are introduced by $\rho = r/a$, $\tau = t/t_0$, with $t_0 = a^2/\alpha$. At the normalized time $\tau \sim 0.05$, the temperature at the axis of the cylinder first begins to decrease from the value of $T_1$, so that subsequent thermal loading of the cylinder is less severe than before. We note that throughout the process, excellent approximation to the temperature distribution can be obtained using power law profiles of the type

$$1 - \frac{T(r,t) - T_0}{T_1 - T_0} = \frac{T_1 - T(r,t)}{T_1 - T_0} = \rho^m. \tag{5.16}$$

In particular, we note that the temperature distribution at $\tau \sim 0.07$ is approximated well by a parabolic profile ($m = 2$). At that moment, the eigenstrain distribution introduced by the inhomogeneous temperature distribution can be described by

$$\varepsilon_r^* = \varepsilon_\theta^* = -\gamma \rho^m. \tag{5.17}$$

Let us determine the stress–strain state that arises in the cylinder during quenching and subsequently upon complete cooling to the ambient temperature.

Assuming the deformation to be axisymmetric and plane strain, we note that it can be fully described by the radial displacement function $u(r)$. The strain equations take the form

$$\varepsilon_r = e_r + \varepsilon_r^* = e_r - \gamma r^m = u_{,r}, \quad \varepsilon_\theta = e_\theta + \varepsilon_\theta^* = e_\theta - \gamma r^m = u/r. \tag{5.18}$$

Therefore, the elastic strains are expressed in terms of the displacement function and the thermal strain as follows:

$$e_r = u_{,r} + \gamma r^m, \quad e_\theta = u/r + \gamma r^m. \tag{5.19}$$

The stresses are given by

$$\sigma_r = \frac{E}{(1+v)(1-2v)}[(1-v)e_r + ve_\theta], \quad \sigma_\theta = \frac{E}{(1+v)(1-2v)}[(1-v)e_\theta + ve_r].$$

$$(5.20)$$

Introducing the notation

$$E' = \frac{E(1-v)}{(1+v)(1-2v)}, \quad v' = \frac{v}{1-v}, \tag{5.21}$$

the plane strain stress–strain relations are written as

$$\sigma_r = E'[e_r + v'e_\theta], \quad \sigma_\theta = E'[e_\theta + v'e_r]. \tag{5.22}$$

Substituting (Eq. 5.17) and (Eq. 5.20) into the equation of equilibrium:

$$\frac{\sigma_r - \sigma_\theta}{r} + \frac{d\sigma_r}{dr} = 0, \tag{5.23}$$

we obtain the governing equation for the radial displacement component:

$$r^2 u'' + ru' - u + (1+v')\gamma m r^{m+1} = 0. \tag{5.24}$$

The general solution of this equation has the form

$$u(r) = Ar + Br^{-1} - \gamma r^{m+1}(1+v')/(m+2). \tag{5.25}$$

Setting constant $B = 0$ ensures that the solution does not have a singularity at $r = 0$. The elastic strains and stresses are found by back-substitution into (Eq. 5.19) and (Eq. 5.22). The unknown constant $A$ is determined from the boundary condition $\sigma_r(c) = -q$, so that finally

$$\sigma_r(r) = -q - \frac{(1-v'^2)}{m+2}E'\gamma(c^m - r^m),$$

$$(5.26)$$

$$\sigma_\theta(r) = -q - \frac{(1-v'^2)}{m+2}E'\gamma(c^m - (m+1)r^m).$$

We note that the solution for thermally induced stresses in a traction-free cylinder is obtained by substituting $q = 0$, $c = a$ into the above equations. In this case the stresses in the cylinder are given by

$$\sigma_r(r) = \frac{(1-v'^2)}{4}E'\gamma(r^2 - a^2), \quad \sigma_\theta(r) = \frac{(1-v'^2)}{4}E'\gamma(3r^2 - a^2). \tag{5.27}$$

We now consider the onset and consequences of yielding, particularly in terms of the residual stress state within the cylinder. Adopting the Tresca yield criterion, and assuming that the out-of-plane stress component is intermediate between the principal plane stresses, $\sigma_\theta > \sigma_z > \sigma_r$, we write the yield condition as

$$\sigma_\theta - \sigma_r = \sigma_Y. \tag{5.28}$$

The equilibrium equation leads to

$$\frac{d\sigma_r}{dr} - \frac{\sigma_Y}{r} = 0, \tag{5.29}$$

with the solution

$$\sigma_r^p = \sigma_Y \log\frac{r}{a}, \quad \sigma_\theta^p = \sigma_Y\left(1 + \log\frac{r}{a}\right), \quad c \le r \le a. \tag{5.30}$$

The location of the plastic zone boundary is then determined from Eq. (5.26) by the requirement that the yield condition be satisfied at the boundary of the elastic region:

$$\sigma_\theta - \sigma_r = \frac{(1 - \nu'^2)}{2} E'\gamma c^2 = \sigma_Y, \quad c^2 = \frac{2\sigma_Y}{E'\gamma(1 - \nu'^2)}. \tag{5.31}$$

The pressure $q$ acting across the plastic zone boundary $r = c$ is obtained from (Eq. 5.30) as

$$q = -\sigma_r^p(c) = \sigma_Y \log\frac{a}{c}. \tag{5.32}$$

The complete stress state upon yielding is given by

$$\sigma_r(r) = \begin{cases} \sigma_Y \log\dfrac{c}{a} - \dfrac{(1 - \nu'^2)}{4} E'\gamma(c^2 - r^2), & r < c \\[2ex] \sigma_Y \log\dfrac{r}{a}, & c < r < a \end{cases} \tag{5.33}$$

$$\sigma_\theta(r) = \begin{cases} \sigma_Y \log\dfrac{c}{a} - \dfrac{(1 - \nu'^2)}{4} E'\gamma(c^2 - 3r^2), & r < c \\[2ex] \sigma_Y\left(1 + \log\dfrac{r}{a}\right), & c < r < a \end{cases} \tag{5.34}$$

When the cylinder cools down to uniform temperature, the stress change is given by the thermoelastic unloading expressed in (Eq. 5.27). The final residual stress state is therefore given by

$$\sigma_r(r) = \begin{cases} \sigma_Y \log\dfrac{c}{a} + \dfrac{(1 - \nu'^2)}{4} E'\gamma(a^2 - c^2), & r < c \\[2ex] \sigma_Y \log\dfrac{r}{a} - \dfrac{(1 - \nu'^2)}{4} E'\gamma(r^2 - a^2), & c < r < a \end{cases} \tag{5.35}$$

$$\sigma_\theta(r) = \begin{cases} \sigma_Y \log\dfrac{c}{a} + \dfrac{\left(1-v'^2\right)}{4} E'\gamma\left(a^2 - c^2\right), & r < c \\[2em] \sigma_Y\left(1 + \log\dfrac{r}{a}\right) - \dfrac{\left(1-v'^2\right)}{4} E'\gamma\left(3r^2 - a^2\right), & c < r < a \end{cases} \tag{5.36}$$

Note that the residual stress within the core region of the cylinder $r < c$ that remained elastic during the quenching process is equibiaxial and uniform, $\sigma_r(r) = \sigma_r(r) = $ const. Note further that the continuity of the residual stress distribution across the plastic zone boundary $r = c$ is ensured by Eq. (5.31). That relationship can be rewritten in the form that introduces the yield parameter $Y$ as follows:

$$\frac{c^2}{a^2} = \frac{2\sigma_Y}{E'\gamma a^2\left(1 - v'^2\right)} = Y. \tag{5.37}$$

Substitution gives an alternative form:

$$\sigma_r(r) = \begin{cases} \dfrac{1}{2}\sigma_Y\left[(\log Y - 1) + \dfrac{1}{Y}\right], & r < c \\[2em] \sigma_Y\left[\log\dfrac{r}{a} - \dfrac{1}{2Y}\left(\dfrac{r^2}{a^2} - 1\right)\right], & c < r < a \end{cases} \tag{5.38}$$

$$\sigma_\theta(r) = \begin{cases} \dfrac{1}{2}\sigma_Y\left[(\log Y - 1) + \dfrac{1}{Y}\right], & r < c \\[2em] \sigma_Y\left[1 + \log\dfrac{r}{a} - \dfrac{1}{2Y}\left(\dfrac{3r^2}{a^2} - 1\right)\right], & c < r < a \end{cases} \tag{5.39}$$

Fig. 5.8 shows the residual stress distribution due to quenching, plotted assuming problem parameter values $a = 1$, $\sigma_Y = 100$ MPa.

Fig. 5.9 represents an illustration of the residual strain distribution in a quenched cylinder calculated using Young's modulus of 70 GPa and Poisson's ratio of 0.3.

An example of experimentally measured quench-induced residual strains is shown in Fig. 10.4. Although the profiles are qualitatively similar to those obtained from the model and illustrated in the plot given previously, some differences are apparent. Plastic deformation during quenching is sensitive to the temperature dependence of the yield stress. The practical difficulty of determining the conditions of residual stress generation during quenching means that often information about the residual stress is available only from samples cooled to room temperature. The determination of the yielding conditions based on the residual stress state evaluated after cooling can proceed as follows.

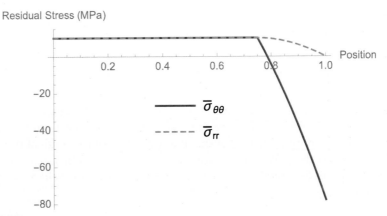

**FIGURE 5.8**
Illustration of residual stress distribution in a quenched cylinder (Korsunsky et al., 2016).

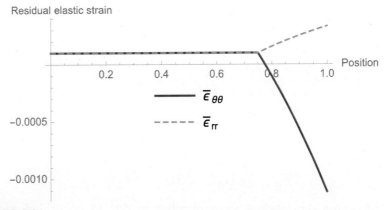

**FIGURE 5.9**
Illustration of residual strain distribution in a quenched cylinder (Korsunsky et al., 2016).

Let the value of $c$ be found from the measured residual stress profile of the kind shown in Fig. 5.8 by locating the extreme radial position on the constant stress plateau. The value of the yielding parameter $Y$ when yielding took place can be found according to Eq. (5.37) from

$$Y = c^2/a^2. \tag{5.40}$$

The equivalent yield stress value at the temperature when plastic flow took place can be calculated by inverting the expression in Eq. (5.39) at $r = a$:

$$\sigma_Y = -\frac{Y\bar{\sigma}_\theta(a)}{(1-Y)}. \tag{5.41}$$

Here $\bar{\sigma}_\theta(a)$ denotes the maximum value of compressive residual stress at the periphery of the cylinder.

Further consideration of the equation reveals a limitation of the aforementioned treatment. The maximum hoop stress $\bar{\sigma}_\theta(a)$ at the periphery reaches the yield stress $\sigma_Y$ for the yield parameter value $Y = 1/2$, i.e., when $c = a/\sqrt{2} \approx 0.707a$. Noting that the yield stress typically decreases with temperature, the substitution of the traction-free condition $\bar{\sigma}_r(a) = 0$ in the Tresca yield criterion $\bar{\sigma}_\theta(a) - \bar{\sigma}_r(a) = \sigma_Y$ indicates that reverse plastic flow is likely to take place during cooling. Consequently, the stress state will change, and Eqs. (5.38) and (5.39) will no longer apply.

# CHAPTER 6

# The Eigenstrain Theory of Residual Stress

*Vladimir Nabokov: Spring in Fialta 1936*

*Andrei Petrov, Taming of the fire 1972*

*Kazimir Malevich, Black suprematist square (oil on canvas, 80×80 cm) 1915, Tretyakov Gallery, Moscow*

## 6.1 GENERALIZATION

Residual stresses play an important role in determining the deformation behavior and fatigue durability of engineering components and assemblies. It is well known, for example, that compressive near-surface residual stresses act to inhibit crack initiation and propagation, and thus affect the fatigue life of the object. On the other hand, residual stresses themselves are known to undergo modification during thermal and mechanical loading, through various mechanisms related to time-independent plasticity, creep, phase transformation, etc.

In previous chapters simple examples of one-dimensional and two-dimensional residual stress distributions were considered, and their relationship with eigenstrains was highlighted. The purpose of this chapter is to provide a generalization of this approach to more complex three-dimensional problems about residual stresses in elastic bodies. In other words, the problem of residual stress is thought of as a generalization of the theory of elasticity with large perturbation in the form of inelastic strain (eigenstrain).

Residual stress states in arbitrarily shaped solid bodies are usually complex, and difficult to describe, because in the general case they must be represented by the six components of the stress tensor varying as a function of three spatial variables. It is virtually impossible to imagine an experimental procedure that would readily and routinely provide this level of detail. At any rate, the interpretation of pointwise data in terms of six independent components is likely to present a serious practical challenge.

**CONTENTS**

**67**

A Teaching Essay on Residual Stresses and Eigenstrains. http://dx.doi.org/10.1016/B978-0-12-810990-8.00006-9

Any residual stress state described by the tensor $\boldsymbol{\sigma}$ must, by definition, be self-equilibrating. This requirement establishes a relationship between the gradients of different components of the stress tensor $\boldsymbol{\sigma}$, expressed as

$$\mathbf{div}\ \boldsymbol{\sigma} = 0. \tag{6.1}$$

Furthermore, the stress state deduced within a residually stressed object must satisfy the traction-free boundary conditions, namely,

$$\boldsymbol{\sigma}\cdot\mathbf{n} = 0, \tag{6.2}$$

where $\mathbf{n}$ denotes the surface normal. However, it is not easy to enforce this requirement on the deduced stress state, or to formulate the constraints that must be imposed on the measured strain data. It is possible, however, to develop a rational analytical approach based on the concept of *eigenstrain* that reduces the size of the data array needed to represent a particular residual stress state, and at the same time guarantees satisfaction of equations of equilibrium (6.1) and traction-free boundary conditions (6.2).

Eigenstrain modeling is a powerful analytical technique for the representation of residual stress states in solids (Mura, 1987). A practical approach to the use of eigenstrain in residual stress modeling (Korsunsky and Withers, 1997; Korsunsky, 2005a,b) has been developed based on the following fundamental postulates:

1. In the absence of eigenstrain (stored inelastic strain) any elastic solid is completely free from residual stress. Indeed, the very definition of elastic material response requires that stresses and strains arise in the body upon the application of an external load, and vanish completely upon load removal.
2. Residual stresses within a solid arise in response to the introduction, through some inelastic mechanism (plasticity, creep, cutting and pasting, phase transformation, etc.) of permanent *nonuniform* strains within the body. Note, however, that the introduction of an entirely spatially uniform permanent strain field does not, in fact, lead to the generation of residual stresses.
3. Elastic and inelastic strains are additive, i.e.,

$$\boldsymbol{\varepsilon} = \boldsymbol{\varepsilon}^* + \mathbf{e}, \text{ or in index notation, } \varepsilon_{ij} = \varepsilon_{ij}^* + e_{ij}, \tag{6.3}$$

where $\varepsilon_{ij}$ denotes the total strain, $e_{ij}$ denotes the elastic strain, and $\varepsilon_{ij}^*$ denotes eigenstrain.

4. Total strain must be compatible, i.e., satisfy

$$\mathrm{rot}\big((\mathrm{rot}\ \varepsilon)^T\big) = 0,$$

leading to relationships between strain components of the type

$$\frac{\partial^2 \varepsilon_{xx}}{\partial y^2} + \frac{\partial^2 \varepsilon_{yy}}{\partial x^2} - 2\frac{\partial^2 \varepsilon_{xy}}{\partial x \partial y} = 0. \tag{6.4}$$

5. Eigenstrains (permanent inelastic strains) act as the sources of incompatibility of displacement, i.e., can be thought of as appearing in the right hand of the Saint-Venant compatibility equations. Indeed, from the compatibility Eq. (6.4) one readily obtains the "incompatibility" equation for the elastic strain, $e_{ij}$, in the following form:

$$\frac{\partial^2 e_{xx}}{\partial y^2} + \frac{\partial^2 e_{yy}}{\partial x^2} - 2\frac{\partial^2 e_{xy}}{\partial x \partial y} = \Xi, \tag{6.5}$$

where

$$\Xi = 2\frac{\partial^2 \varepsilon^*_{xy}}{\partial x \partial y} - \frac{\partial^2 \varepsilon^*_{xx}}{\partial y^2} - \frac{\partial^2 \varepsilon^*_{yy}}{\partial x^2}. \tag{6.6}$$

Note from the above expression that the "forcing term," $\Xi$, turns to zero for uniform eigenstrains. In fact, it also vanishes for eigenstrains that depend linearly on one coordinate and do not depend at all on any other coordinates, predicting that bodies containing such residual strain distributions are also free from residual elastic strain, and hence from residual stress.

6. The problem of determining the residual elastic fields (residual elastic strain and residual stresses, as well as residual deformations, i.e., distortions) due to the given eigenstrain, $\varepsilon^*_{ij}$, requires simultaneous solution of Eqs. (6.1), (6.2), and (6.5), together with the constitutive law in the form of elasticity equations (generalized Hooke's law):

$$\boldsymbol{\sigma} = \mathbf{C} : \mathbf{e}, \text{ or in index notation, } \sigma_{kl} = C_{klij} e_{kl}. \tag{6.7}$$

7. The eigenstrain problem is not in fact in any way different from the well-known thermoelastic problem, in which the forcing term $\Xi$ arises as a result of thermal gradients (note that an unconstrained uniformly heated body remains stress-free). In fact, arbitrary eigenstrain distributions can be successfully simulated by means of anisotropic thermal expansion.

Mura (1987) presents an analytical framework for eigenstrain analysis based on the Green's function for elasticity. Green's function $G_{km}(\mathbf{x} - \mathbf{x}')$ expresses the equilibrium in terms of the displacement $u_k$ at $\mathbf{x}$ due to the unit point body force in direction $m$ applied at $\mathbf{x}'$, and satisfies the following fundamental equation:

$$C_{ijkl} G_{km,lj}(\mathbf{x} - \mathbf{x}') = -\delta_{im}\delta(\mathbf{x} - \mathbf{x}'). \tag{6.8}$$

For isotropic linear elastic body

$$G_{km}(\mathbf{x}) = \frac{1}{16\pi\mu(1-\nu)x}\left[(3-4\nu)\delta_{ij} - \frac{x_i x_j}{x^2}\right].$$ (6.9)

For an arbitrary distribution of eigenstrain the displacement field is obtained by convolving the distribution with the appropriate derivative of Green's function:

$$u_i(\mathbf{x}) = -\int_{-\infty}^{\infty} C_{jlmn}\varepsilon^*_{mn}G_{ij,l}(\mathbf{x}-\mathbf{x}')d\mathbf{x}'.$$ (6.10)

Thus any displacement (and hence strain and stress) fields due to a given eigenstrain distribution can be computed by integration.

## 6.2 THE EIGENSTRAIN CYLINDER

The problem of an infinitely extended right circular cylinder that undergoes a constant dilatational inelastic strain provides a convenient introduction to the solution of residual stress problems using the eigenstrain approach. Consider an infinitely extended elastic solid in which a domain $\{\Omega\text{:}r \leq a\}$ is defined in cylindrical polar coordinates. The geometry of the problem is illustrated in Fig. 6.1 for $a = 1$. The problem is formulated in terms of letting the

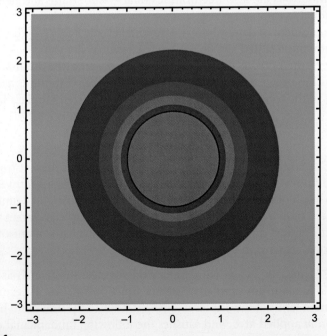

**FIGURE 6.1**
The eigenstrain cylinder.

material within $\Omega$ experience eigenstrain $\varepsilon^* = \varepsilon^*_{xx} = \varepsilon^*_{yy}$, and determining the elastic fields everywhere within the solid.

The solution proceeds by determining the stress–strain fields within the domain $\Omega$ and outside it, and imposing the requirements of strain compatibility between the two regions to obtain the solution. Within $\Omega$ the stress state is described by uniform stresses ($\sigma_r = \sigma_\theta = -p, \sigma_z$). Assuming the problem to be plane strain, the axial elastic strain is written:

$$e_z = \frac{1}{E}[\sigma_z - v(\sigma_r + \sigma_\theta)] = \frac{1}{E}[\sigma_z + 2vp] = 0. \qquad (6.11)$$

Hence $\sigma_z = -2vp$, and the elastic strain in $\Omega$ is

$$e_\theta = e_r = \frac{1}{E}[\sigma_r - v(\sigma_\theta + \sigma_z)] = \frac{1}{E}[-p + vp + 2v^2 p] = -\frac{p}{E}(1 + v)(1 - 2v). \qquad (6.12)$$

The total strain in $\Omega$ is

$$\varepsilon_\theta = \varepsilon_r = \varepsilon^* - \frac{p}{E}(1 + v)(1 - 2v), \qquad (6.13)$$

hence the radial displacement $u$ in $\Omega$ is given by

$$\frac{u}{r} = \varepsilon_\theta = \varepsilon^* - \frac{p}{E}(1 + v)(1 - 2v), \qquad (6.14)$$

and, in particular, at the boundary $r = a$ of $\Omega$:

$$\frac{u(a)}{a} = \varepsilon_\theta = \varepsilon^* - \frac{p}{E}(1 + v)(1 - 2v). \qquad (6.15)$$

Now consider the stress field in the exterior of domain $\Omega$: it is given by

$$\sigma_r = -p\frac{a^2}{r^2}, \quad \sigma_\theta = p\frac{a^2}{r^2}. \qquad (6.16)$$

The plane strain assumption gives

$$e_z = \frac{1}{E}[\sigma_z - v(\sigma_r + \sigma_\theta)] = \frac{1}{E}\sigma_z = 0, \qquad (6.17)$$

so that $\sigma_z = 0$, and the elastic and total hoop strains in the exterior domain are

$$\varepsilon_\theta = e_\theta = \frac{1}{E}[\sigma_\theta - v(\sigma_r + \sigma_z)] = \frac{p}{E}\frac{a^2}{r^2}(1 + v). \qquad (6.18)$$

The radial displacement at $r = a$ in the exterior domain is therefore

$$\frac{u(a)}{a} = \frac{p}{E}(1 + v). \qquad (6.19)$$

Strain compatibility (material continuity) in this simple case reduces to combining Eqs. (6.15) and (6.19) to obtain

$$\varepsilon^* - \frac{p}{E}(1+\nu)(1-2\nu) = \frac{p}{E}(1+\nu), \tag{6.20}$$

from which it follows that $E\varepsilon^* = 2p(1-\nu^2)$ and

$$p = \frac{E\varepsilon^*}{2(1-\nu^2)}. \tag{6.21}$$

This completes the solution: the pressure in the eigenstrain cylinder has been found, and the elastic fields given by Eqs. (6.12), (6.14), (6.16), and (6.17) are all expressed in terms of the eigenstrain $\varepsilon^*$.

It is worth discussing this example in some more detail, because the key steps of the solution are immediately apparent because of their simplicity. This simplicity in part flows from the fact that eigenstrain is uniform in domain $\Omega$, and equals zero outside it. Therefore, the elastic fields in these domains that satisfy stress equilibrium and strain compatibility conditions can be written readily, and the problem solution is accomplished by matching the tractions, strains, and displacements across the domain boundary $\partial\Omega$.

## 6.3   THE EIGENSTRAIN SPHERE

As another example consider the case of a three-dimensional infinitely extended elastic solid described by spherical coordinates $(R,\vartheta,\varphi)$. We define a spherical domain $\Omega$ of radius $a$ that is separated from the exterior by its boundary $\{\partial\Omega : R = a\}$. We consider uniform expansion of the spherical domain $\Omega$ that is described by the eigenstrain $\varepsilon_r^* = \varepsilon_\vartheta^* = \varepsilon_\varphi^* = \varepsilon^*$. In particular it may be considered that the strain misfit represented by this eigenstrain arises because of a temperature change $\Delta T$ and a mismatch in the coefficient of linear thermal expansion $\Delta\alpha$ between the interior and exterior domains: $\varepsilon^* = \Delta\alpha\Delta T$.

The spherical eigenstrain domain $\Omega$ isotropically compressed by pressure $p$ experiences stresses $\sigma_R = \sigma_\vartheta = \sigma_\varphi = -p$, elastic strains $e_r = e_\vartheta = e_\varphi = -\frac{p}{E}(1-2\nu)$, and hence the total circumferential strain at the interior boundary $R = a$ given by

$$\frac{u(a)}{a} = \varepsilon_\vartheta = \varepsilon^* - \frac{p}{E}(1-2\nu). \tag{6.22}$$

According to Timoshenko and Goodier (1951), the radial stress in the exterior of an internally pressurized spherical cavity in an infinitely extended elastic solid is given by

$$\sigma_R = -p\frac{a^3}{R^3}, \quad \sigma_\vartheta = \sigma_\varphi = \sigma_R + \frac{R}{2}\frac{\partial\sigma_R}{\partial R} = p\frac{a^3}{2R^3}. \qquad (6.23)$$

The circumferential strain is given by

$$\varepsilon_\vartheta = \frac{1}{E}\left[\sigma_\vartheta - \nu(\sigma_R + \sigma_\varphi)\right] = \frac{p}{E}\frac{a^3}{2R^3}(1+\nu) \qquad (6.24)$$

and evaluating it at the exterior boundary gives

$$\frac{u(a)}{a} = \varepsilon_\vartheta|_{R=a} = \frac{p}{2E}(1+\nu). \qquad (6.25)$$

Ensuring compatibility requires equating (6.22) and (6.25), resulting in the following relationship:

$$\varepsilon^* = \frac{3p}{2E}(1-\nu), \quad p = \frac{2}{3}\frac{E\varepsilon^*}{(1-\nu)}. \qquad (6.26)$$

We note in particular the expressions for displacements, strains, and stresses in the exterior of domain $\Omega$, $r > a$:

$$u(R) = \frac{pa^3}{2ER^2}(1+\nu), \quad \varepsilon_\vartheta = \frac{pa^3}{2ER^3}(1+\nu), \quad \sigma_R = -\frac{pa^3}{R^3}, \quad \sigma_\vartheta = \sigma_\varphi = \frac{pa^3}{2R^3}. \qquad (6.27)$$

In fact, the treatment of solutions for a spherical cavity based on the Love's stress function presented by Timoshenko and Goodier (1951) admits a much more flexible family of solutions. As an example, we seek to find the solutions for two different kinds of eigenstrain: (1) axial, given by $\varepsilon_z^* = \varepsilon^*$, $\varepsilon_r^* = \varepsilon_\theta^* = 0$, and (2) transversely isotropic, given by $\varepsilon_r^* = \varepsilon_\theta^* = \varepsilon^*$, $\varepsilon_z^* = 0$. It is worth noting immediately that the sum of the two solutions (1) + (2) with equal eigenstrains once again gives rise to the solution for an isotropically expanded sphere.

## 6.4 ESHELBY ELLIPSOIDAL INCLUSIONS

The celebrated Eshelby (1957) solution has served as the basis for numerous studies in micromechanics of precipitates, polycrystals, cracks, voids, porous materials and composites, etc. The significance of this solution for the present discussion lies in the fact that it represents an important instance of residual stress state due to the introduction of arbitrary uniform eigenstrain within a finite domain.

The closed-form solution provides the expressions for the elastic fields inside and outside an ellipsoidal inclusion defined as domain $\Omega$ in the Cartesian coordinates $(x_1, x_2, x_3)$ given by

$$\frac{x_1^2}{a_1^2} + \frac{x_2^2}{a_2^2} + \frac{x_3^2}{a_3^2} \leq 1, \tag{6.28}$$

where the semiaxes are usually numbered so that $a_1 \geq a_2 \geq a_3$. Eshelby considered the introduction of arbitrary *uniform* eigenstrain $\varepsilon_{kl}^*$ within $\Omega$, and derived the general form of the interior and exterior elastic solutions for the general case of arbitrary material elastic anisotropy. In the simple case of isotropic material, for the inclusions in the form of a sphere ($a_1 = a_2 = a_3$) and of a right circular cylinder ($a_1 \to \infty$, $a_2 = a_3$), the solution for the elastic strains within the inclusion was uniform. By proving that this is the case for any ellipsoidal shape and any material, Eshelby achieved a very important reduction in complexity of the problem, namely, showing that the relationship between the elastic strains $e_{ij}$ within the inclusion that arise as a result of the introduction of eigenstrain $\varepsilon_{kl}^*$ is linear and can be written in terms of a *constant* tensor:

$$e_{ij} = S_{ijkl}\varepsilon_{kl}^*, \quad x \in \Omega \tag{6.29}$$

where the Eshelby tensor $S_{ijkl} = S_{jikl} = S_{ijlk}$ possesses minor (but not major) indicial symmetries and depends only on the ellipsoid semiaxes and material's Poisson's ratio.

In contrast, the exterior elastic field depends on the coordinate x:

$$e_{ij}(x) = D_{ijkl}(x)\varepsilon_{kl}^*, \quad x \notin \Omega. \tag{6.30}$$

The interior Eshelby solution for an isotropic material is given by (Mura, 1987):

$$S_{1111} = \frac{3}{8\pi(1-v)} a_1^2 I_{11} + \frac{(1-2v)}{8\pi(1-v)} I_1,$$

$$S_{1122} = \frac{1}{8\pi(1-v)} a_2^2 I_{12} - \frac{(1-2v)}{8\pi(1-v)} I_1;$$

$$\tag{6.31}$$

$$S_{1133} = \frac{1}{8\pi(1-v)} a_3^2 I_{13} - \frac{(1-2v)}{8\pi(1-v)} I_1,$$

$$S_{1212} = \frac{a_1^2 + a_2^2}{16\pi(1-v)} I_{12} - \frac{(1-2v)}{16\pi(1-v)(I_1+I_2)}.$$

here $I_i$, $I_{ij}$ are integral expressions that have an implied dependence on the parameter $\lambda$ that is equal to the largest positive root of the equation:

$$\frac{x_1^2}{a_1^2 + \lambda} + \frac{x_2^2}{a_2^2 + \lambda} + \frac{x_3^2}{a_3^2 + \lambda} = 1. \tag{6.32}$$

A positive root always exists for points in the exterior domain ($x \notin \Omega$), and becomes equal to zero on the domain boundary $\partial\Omega$, as is evident from its definition in Eq. (6.28). Positive root does not exist for points inside the ellipsoid, and $\lambda = 0$ should be assumed for interior points. For the purposes of calculation it may be convenient to think of the parameter $\lambda$ as a continuous single-valued real function of the field point and ellipsoid geometry, $\lambda = \lambda(x, a)$.

The integrals $I_i$, $I_{ij}$ that appear in the definition of the interior and exterior Eshelby tensors are given by

$$I_1(\lambda) = 2\pi a_1 a_2 a_3 \int_\lambda^\infty \frac{ds}{(a_1^2 + s)\Delta(s)},$$

$$I_{11}(\lambda) = 2\pi a_1 a_2 a_3 \int_\lambda^\infty \frac{ds}{(a_1^2 + s)^2 \Delta(s)}, \qquad (6.33)$$

$$I_{12}(\lambda) = 2\pi a_1 a_2 a_3 \int_\lambda^\infty \frac{ds}{(a_1^2 + s)(a_2^2 + s)\Delta(s)},$$

where $\Delta(s) = (a_1^2 + s)^{1/2}(a_2^2 + s)^{1/2}(a_3^2 + s)^{1/2}$, and the integrals for all other indices are obtained by cyclic permutations.

For the purposes of computation, all $I$-integrals can be expressed in terms of elliptic functions $E(\theta, k)$ and $F(\theta, k)$:

$$F(\theta, k) = \int_0^\theta \left(1 - k^2 \sin^2 t\right)^{-1/2} dt, \quad E(\theta, k) = \int_0^\theta \left(1 - k^2 \sin^2 t\right)^{1/2} dt. \qquad (6.34)$$

where parameters $\theta, k$ are defined as

$$\theta(\lambda) = \arcsin\left(\frac{a_1^2 - a_3^2}{a_1^2 + \lambda}\right)^{1/2}, \quad k = \left(\frac{a_1^2 - a_2^2}{a_1^2 - a_3^2}\right)^{1/2}. \qquad (6.35)$$

Eshelby inclusion ellipsoidal solution offers remarkable conceptual clarity and simplicity. Furthermore, its parametric nature means that the geometry of the inclusion and its elastic properties mismatch with the matrix can be adjusted to serve as an approximation for a great variety of situations encountered in materials science: penny-shaped and slit-like cracks, rigid plate-like inclusions and fibers, precipitations of second phase that possess different volume and stiffness, etc. Owing to its versatility, the Eshelby solution has been used as the basis for various composite and polycrystal models: self-consistent, Mori-Tanaka, etc.

However, owing to the complex nature of the analytical expressions for the exterior fields, further use of the solution has been somewhat restricted. The nature of the computation requires analytical manipulations to be performed first, and then incorporated in numerical evaluation to obtain strain and stress maps. This type of calculation can be efficiently accomplished using a symbolic algebra package such as Mathematica. Nevertheless, to the author's surprise no such package seems to be available. To aid in the progress toward such package, the expressions for *I*-integrals and their derivatives used in the computation of the Eshelby solution are given in Appendix A.

> **Question 6.1**: Write a notebook in Mathematica to evaluate exterior strain and stress fields outside an ellipsoidal inclusion of arbitrary shape containing arbitrary uniform *eigenstrain* state.

## 6.5   NUCLEI OF STRAIN

Consider Eq. (6.19) and take the limit of the vanishing radius of the spherical cavity $a \to 0$, while increasing the internal pressure, so that $pa^3 = C$:

$$u(R) = \frac{C}{2ER^2}(1+\nu), \quad \varepsilon_\vartheta = \frac{C}{2ER^3}(1+\nu), \quad \sigma_R = -\frac{C}{R^3}, \quad \sigma_\vartheta = \sigma_\varphi = \frac{C}{2R^3}. \quad (6.36)$$

We obtained the solution for a *center of expansion*. It can also be thought of as an instance of dilatational eigenstrain concentrated at a point. The result presents an example of a *nucleus of strain*, a fundamental singular solution in the theory of elasticity that gives rise to a particular distribution of residual stresses and residual elastic strains.

Some reference to the family of fundamental singular solutions in the theory of elasticity, or *strain nuclei*, has already been made in previous chapters. They play the role of "charges" in the theory of elasticity, i.e., provide singular sources of deformation in the same way as charges act as point sources of electrostatic field. In the same ways as charges give rise to electrostatic fields, strain nuclei give rise to self-equilibrating residual stress fields, which explains their significance for residual stress analysis. An important difference between the two theories arises because the fundamental field quantities in elasticity are second rank tensors, as opposed to the vector (first rank tensor) quantities that describe the electric (and magnetic) fields. As a consequence, the more complex theory of elasticity gives rise to a richer family of fundamental solutions that merits careful study.

The common *nuclei of strain* that act as sources of inelastic strains are:

- point defects: vacancies, interstitials, substitutional atoms

- line defects (dislocations) and their arrays, e.g., low angle grain boundaries
- planar defects: dislocation loops, stacking faults, twins
- volumetric defects associated with transformation strains, e.g., martensitic laths, precipitates

Dislocations play a particularly important role: not only do they act as residual stress sources at the mesoscale (i.e., at a scale between atomic and macroscopic), but they also respond to internal stresses that act upon them by glide or climb, two modes of motion that at moderate temperatures are distinct as being "easy" and "hard," in that the former requires no long distance mass transport and can be accomplished by minimal local change in the atomic positions, and is known to be the principal mechanism of time-independent metal plasticity, whereas the latter relies on thermal activation, and hence becomes progressively more significant at higher temperatures and mediates creep. This means that the migration of dislocations under the action of internal stresses can be used to simulate the evolution of inelastic strain (and residual stress). This approach has become known as discrete dislocation dynamics and has helped obtain conceptual insights into a variety of interesting small-scale phenomena in materials mechanics. Dislocation dynamics modeling has enabled simulating strain gradient effects associated with grain boundaries and interfaces.

In addition, dislocations correspond to displacement discontinuity that makes them eminently suitable for modeling cracks. This has motivated the development of distributed or continuous dislocation modeling of cracks (Hills et al., 1996).

In view of the significance of dislocations as nuclei of strain responsible for residual stress and its evolution, the next chapter is devoted to this topic.

# CHAPTER 7

# Dislocations

*Irwin Shaw: The girls in their summer dresses. Mixed Company 1950*

*Dmitri Shostakovich, Valse from Suite no. 2 for Jazz Orchestra 1938*

*Jan Vermeer van Delft, The Girl with a Pearl Earring (oil on canvas, 47 × 40 cm) 1666*
*Mauritshuis, The Hague*

## 7.1 DISLOCATIONS

Dislocations play a particularly important role in the theory of residual stresses for two reasons:

1. They are a convenient elementary source of eigenstrain
2. They move under applied shear stress by glide and thus are responsible for plastic deformation (at constant volume) and eigenstrain generation

Dislocations are associated with crystal lattices. The nature (intensity and orientation) of a dislocation is characterized by the Burgers' vector that is the closure of the circuit around the dislocation line through the crystal lattice. Depending on the orientation relationship between Burgers' vector and the tangent to the dislocation line, segments of dislocation lines can be classified into screw (parallel) and edge (normal).

Dislocations correspond to one of the fundamental cases of localized inelastic strain distributions (*nuclei of strain*) and thus not only serve as an example of the most important class of crystal defects, but also provide very a useful mathematical tool for the analysis of residual stresses.

## 7.2 SCREW DISLOCATION: ANTIPLANE SHEAR SOLUTION

Consider deformation such that the only nonzero elastic fields are:

displacement $u_3(x,y)$, strains $\varepsilon_{13}(x,y)$, $\varepsilon_{23}(x,y)$, stress $\sigma_{13}(x,y)$, $\sigma_{23}(x,y)$.

### CONTENTS

**79**

A Teaching Essay on Residual Stresses and Eigenstrains. http://dx.doi.org/10.1016/B978-0-12-810990-8.00007-0

Conditions of strain compatibility are satisfied by construction.

Equations of equilibrium reduce to a single Laplacian equation:

$$\Delta u_3(x) = 0. \tag{7.1}$$

A general family of solutions of this type can therefore be found by considering all functions of two variables that are harmonic:

$$u_3(x, y) = \varphi(x, y), \quad \text{where} \quad \Delta\varphi(x, y) = 0. \tag{7.2}$$

Consider eigenstrain prescribed as follows:

$$\varepsilon_{23}^*(x) = \frac{1}{2}b\delta(y)H(-x). \tag{7.3}$$

all other components being zero everywhere.

Burgers showed that the corresponding deformation field is antiplane strain, i.e., the only displacement component is given by

$$u_3(x) = \frac{1}{2\pi}b\arctan(y/x). \tag{7.4}$$

The nonzero stress components are illustrated in Fig. 7.1 and are given by

$$\sigma_{31} = \mu u_{3,1} = -\frac{\mu b}{2\pi}\frac{y}{(x^2+y^2)},$$

$$\tag{7.5}$$

$$\sigma_{32} = \mu u_{3,2} = -\frac{\mu b}{2\pi}\frac{x}{(x^2+y^2)}.$$

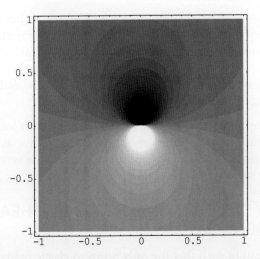

**FIGURE 7.1**
Residual stress $\sigma_{31} = \mu u_{3,1} = -\frac{\mu b}{2\pi}\frac{y}{(x^2+y^2)}$ around a screw dislocation.

## 7.3   EDGE DISLOCATION: PLANE STRAIN SOLUTION

Consider deformation such that displacements are confined to a plane:

displacements $u_1(x,y)$, $u_2(x,y)$,
strains $\varepsilon_{11}(x,y)$, $\varepsilon_{22}(x,y)$, $\varepsilon_{12}(x,y)$,
stresses $\sigma_{11}(x,y)$, $\sigma_{22}(x,y)$, $\sigma_{12}(x,y)$.

The stress state can be described by the Airy stress function:

$$\sigma_{11}(x,y) = \varphi_{,yy}(x,y), \quad \sigma_{22}(x,y) = \varphi_{,xx}(x,y), \quad \sigma_{12}(x,y) = -\varphi_{,xy}(x,y). \tag{7.6}$$

The conditions of strain compatibility impose the requirement that the Airy stress function is *biharmonic*:

$$\Delta\Delta\varphi(x,y) = 0. \tag{7.7}$$

Now the general solution of plane problem of elasticity can be sought in terms of all the solutions of the biharmonic equation.

Consider eigenstrain prescribed as follows:

$$\varepsilon^*21(\mathbf{X}) = \frac{1}{2}b\delta(y)H(-x). \tag{7.8}$$

all other components being zero everywhere.

The corresponding deformation is plane strain. The displacement components are

$$u_1(\mathbf{X}) = \frac{b}{2\pi}\arctan(y/x) + \frac{b}{4\pi(1-v)}\frac{xy}{x^2+y^2},$$

$$\tag{7.9}$$

$$u_2(\mathbf{X}) = \frac{b(2v-1)}{8\pi(1-v)}\log(x^2+y^2) + \frac{b}{4\pi(1-v)}\frac{y^2}{x^2+y^2}.$$

The nonzero stress components are $\sigma_{11}(x,y)$, $\sigma_{22}(x,y)$, $\sigma_{12}(x,y)$, e.g.,

$$\sigma_{12}(\mathbf{X}) = \frac{\mu b}{2\pi(1-v)}\frac{x(x^2-y^2)}{(x^2+y^2)^2}. \tag{7.10}$$

The stress field corresponds to the Airy stress function:

$$\varphi(x,y) = -\mu b y \log\sqrt{x^2+y^2}. \tag{7.11}$$

Fig. 7.2 illustrates the distributions of residual stresses $\sigma_{11}(x,y)$, $\sigma_{22}(x,y)$, $\sigma_{12}(x,y)$ around an edge dislocation.

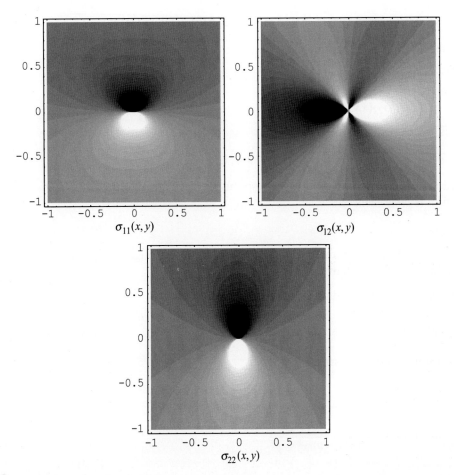

**FIGURE 7.2**

Residual stresses $\sigma_{11}(x,y)$, $\sigma_{22}(x,y)$, $\sigma_{12}(x,y)$ around an edge dislocation.

## 7.4 DISLOCATION PHENOMENOLOGY: FORCES, DIPOLES, INITIATION, AND ANNIHILATION

Point dislocations are an idealization used in plane problem formulations confined to a plane perpendicular to the dislocation line. Real dislocations are extended lines along which the crystal structure has been disturbed. Under an applied external stress $\boldsymbol{\sigma}$ (including stress arising from other dislocations), an element of dislocation line with the Burgers' vector $\mathbf{b}$ and tangent $\mathbf{t}$ experiences the *Peach-Koehler force* (Fig. 7.3):

$$\mathbf{f} = \mathbf{t} \times (\boldsymbol{\sigma} \cdot \mathbf{b}). \tag{7.12}$$

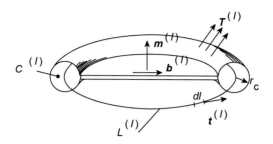

**FIGURE 7.3**
Illustration for the definition of Peach-Koehler force.

Its projection of this force on the glide direction given by the cross-product
$\mathbf{t} \times \mathbf{m}$ is given by

$$f = \mathbf{m} \cdot (\boldsymbol{\sigma} \cdot \mathbf{b}). \tag{7.13}$$

Dislocations move not only in response to external shear stresses, but also because of the elastic stresses generated by other dislocations. From the analysis of stress fields around dislocations shown earlier it is possible to deduce whether dislocations in certain relative positions attract or repel each other (Fig. 7.4).

Dislocations of the same sign on the same slip plane repel, and of opposite signs attract. Dislocations on different planes attract each other and create dislocation walls, or low-angle boundaries.

Dislocations interact and often become organized in groups. The most elementary grouping of dislocations is a dipole: a pair of dislocations of opposite signs. However, dislocations do not cancel each other unless they are exactly in the same position.

If the two dislocations lie in the same glide plane, the dipole could be termed a "glide dipole." If the two dislocations are opposite each other in different glide planes, the dipole is termed "prismatic dipole." Fig. 7.5 illustrates the residual stress fields around a finite prismatic dipole.

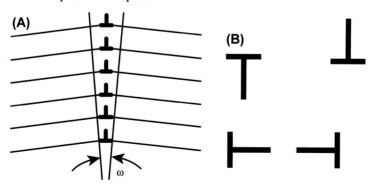

**FIGURE 7.4**
(A) A low-angle grain boundary represented by an array of edge dislocations and (B) schematic notation for the glide and prismatic dipoles.

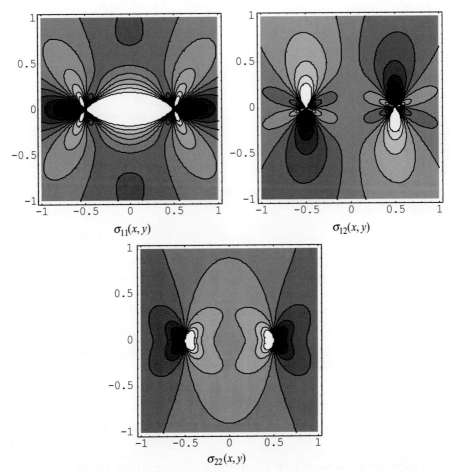

**FIGURE 7.5**

Residual stresses $\sigma_{11}(x,y)$, $\sigma_{12}(x,y)$, $\sigma_{22}(x,y)$ around a finite edge dislocation dipole.

The two dislocations within a glide dipole are attracted to each other and can annihilate. On the other hand, under the action of an external shear the dipole would expand. It is possible to consider the process of creation of a glide dipole from virgin undisturbed crystal under the action of external shear stress.

Imagine dipoles of different lengths being created spontaneously (e.g., due to thermal fluctuations). The attractive force on a dislocation in a glide dipole of length $L_{\text{nuc}}$ must be balanced by the force due to the externally applied shear:

$$F = \frac{\mu b}{2\pi(1-v)} \frac{b}{L_{nuc}}. \tag{7.14}$$

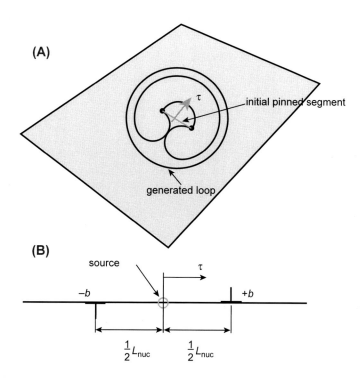

**FIGURE 7.6**
The Frank–Read source.

Hence the shear strength of a dipole is related to the nucleation length by

$$\tau_{nuc} = \frac{\mu b}{2\pi(1-v)L_{nuc}}. \tag{7.15}$$

This equation applies to the popular three-dimensional model of dislocation dipole generation from a pair of "pinning" defects is the Frank-Read source (Fig. 7.6).

## 7.5 THE DYNAMICS OF DISLOCATION MOTION

Dislocations provide an easy mechanism for creating permanent deformation of the crystal lattice by shear. A frequently used analogy is that of the movement of a caterpillar or of creases in a carpet (Fig. 7.7).

Plastic strain due to dislocation movement can be readily estimated. When a single dislocation glides a distance $x$ across a crystal of width $L$ and height $h$, the displacement of one side of the crystal with respect to the other is $u = b$ $x/L$. The permanent plastic shear strain due to $N$ dislocations is computed as $Nbx/(Lh) = \rho bx$, where $\rho$ is the (area) density of mobile dislocations.

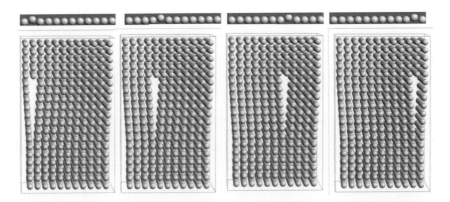

**FIGURE 7.7**
An analogy between the progress of a dislocation by slip across a crystal (lower sequence of images) and the advance of a "caterpillar" effected by maintaining attachment to the ground at all but one point along its length (upper sequence of images). *Based on the original analogy drawn by Orowan, E., in Chapter 3 of "Dislocations in Metals", by Koehler, J.S., Seitz, F., Read, W.T., Shockley, W., Orowan, E., 1954. Am. Inst. Mining and Metall. Engineers, New York.*

Dislocations are a source of eigenstrain, or inelastic deformation. Dislocations "carry" plasticity, similarly to electrical charges in electromagnetism. Dislocations create plastic deformation by moving along their glide planes in response to externally applied stresses. This motion is (1) easy compared with the shift of entire lattice plane at once, and (2) conserves material volume.

As far as the dislocation dynamics is concerned, dislocations (1) exert elastic fields and (2) respond to applied stress by gliding along the slip plane. These properties provide a framework for describing dislocation interaction and thus modeling the process of plastic deformation and its consequences (residual stresses).

Dislocation motion along a glide plane could be thought to obey Newtonian dynamics:

$$m\dot{v} = f - f_{resist}. \tag{7.16}$$

However, we may choose to neglect the inertia (mass) of dislocations and assume quasistatic motion so that equilibrium is always maintained, which requires

$$f = f_{resist}. \tag{7.17}$$

We can next assume that due to phonon emission (lattice vibration), etc., the resisting force on dislocation is a drag force of the form

$$f_{resist} = Bv. \tag{7.18}$$

It follows that dislocation motion can be described simply as

$$v = (\tau + \tilde{\tau})b/B. \tag{7.19}$$

Here $\tau$ refers to the appropriate shear component of the applied stress, and the term $\tilde{\tau}$ represents the influence of other dislocations added to the externally applied stress, i.e., dislocation interaction.

## 7.6 DISLOCATION DYNAMICS EXAMPLES

Beam bending once again provides a convenient starting point for the discussion of the simulation of permanent plastic deformation and residual stresses using dislocation dynamics.

Let there be $N$ dislocations with Burgers' vectors $b$ distributed within a residually bent beam (Fig. 7.8). The total deficit of displacement to one side will be $Nb$, angle $\theta$ of bending $Nb/h$, and curvature $k = Nb/hL = \rho b$. Hence the expression for bent beam curvature is given by $\rho = k/b$, where $\rho = N/hL$ is used to denote the density of *geometrically necessary* dislocations. This term (due to Ashby) is reserved for dislocations that collectively result in the creation of strain gradient, which, as we note from previous discussions, is an attendant feature of beam bending. The creation of a plastic strain gradient requires preponderance of dislocations of a particular sign, as in the earlier illustration. In contrast, *statistically stored dislocations* is a term that refers to dislocations of random sign that can accommodate plastic stretching or compression, but does not give rise to strain gradient.

Fig. 7.9 illustrates the dislocation distribution obtained in numerical simulations of beam bending by van der Giessen et al. (2001). Gray circles represent preinstalled sources of dislocation dipoles. Dislocation pairs could glide on any of the three families of glide planes lying horizontally or at the angles of $\pm 60°$ to the horizontal.

Further illustration of the results of dislocation dynamics simulations carried out by van der Giessen et al. are provided in Figs. 7.10 and 7.11.

Fig. 7.10 shows a snapshot of the dislocation pattern and residual stress state in the vicinity of the crack tip in an imaginary two-dimensional crystal having only two slip planes lying at $\pm 60°$ to the horizontal. Note that the displacement fields associated with the dislocation solutions are used to determine the deformed shape of the crack faces. A region of residual tensile stress appears to persist in front of the crack tip.

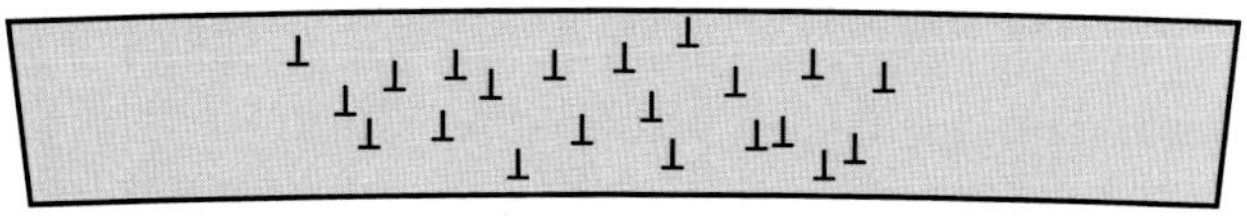

**FIGURE 7.8**
Modeling beam bending using a distribution of $N$ edge dislocations (Van der Giessen et al., 2001).

**(A)**

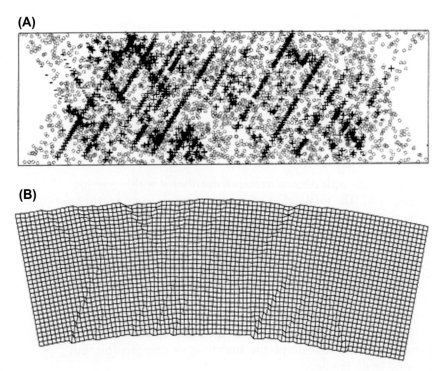

**(B)**

**FIGURE 7.9**

Modeling beam bending using two-dimensional dislocation dynamics: (1) dislocation population and (2) grid distortion (Cleveringa et al., 2000).

Fig. 7.11 illustrates a residual stress component parallel to the interface between a similar hypothetical two-dimensional crystal and an elastic substrate. Note the inhomogeneity of stress in the direction perpendicular to the interface is concentrated within a band. The width of this band is related to the inherent length scale associated with the dislocations.

The power of distributed dislocation dynamics (DDD) to bridge the scales is apparent from these examples, in that they reveal the details of dislocation activity at stress concentrating features, such as crack and notches, or interfaces. Furthermore, they highlight the development of strain gradient features (see near interface band of high stress in Fig. 7.11) that cannot be predicted by classical elastic–plastic modeling: the thickness of the stress concentration band is related to the inverse of the dislocation density, which, importantly, is not a length dimension characteristic for the material, but rather depends on the processing and deformation history.

Nevertheless, discrete dislocation dynamics also has some limitations that arise from the crudeness of the approximation used to describe the nature of these

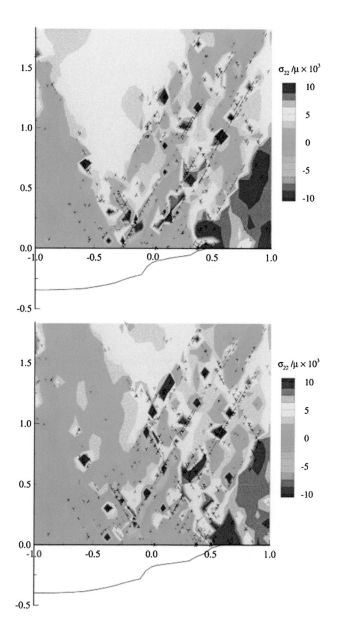

**FIGURE 7.10**

Distribution of the crack-opening stress in the vertical direction, near the tip of a crack in a crystal having two slip systems with slip planes oriented at $\pm 60°$ with respect to the horizontal crack plane. The curve below the horizontal axis gives the opening profile (multiplied by a factor of 10). *From Cleveringa et al. (2000).*

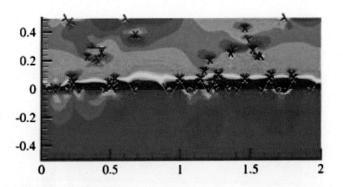

**FIGURE 7.11**

A distribution of horizontal stress parallel to the interface in a single crystal metallic film of 0.5µm thickness resting on a stiff elastic substrate. The film can deform by slip on three planes at 0° and ±60° with respect to the horizontal interface. *From Nicola, L., Van der Giessen, E., Needleman, A., 2003. Discrete dislocation analysis of size effects in thin films. J. Appl. Phys. 93 (10), 5920–5928.*

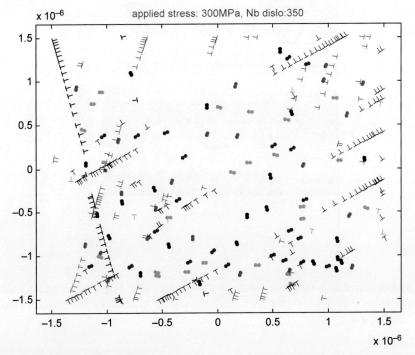

**FIGURE 7.12**

Snapshot of a DDD simulation (Gaucherin et al., 2009).

effects and the crystal in which they exist. For example, the vast majority of real metallic crystals possess significant elastic anisotropy (with the notable exception of tungsten, which has the Zener anisotropy factor of 1.01), meaning that the associated stress and strain fields must be computed accordingly to capture the details of dislocation interaction. Furthermore, slip planes in real crystals do not correspond to the geometry used in simplified formulation use in Figs. 7.9 and 7.10. More realistic simulation requires proper account to be taken of the crystal's anisotropic elastic properties, the three-dimensional nature of dislocation lines and their energies (that affect line bending), the dependence of resistance to glide on dislocation splitting, etc. The literature on this subject is vast and demonstrates that further insight into the details of plastic deformation of metals can be gained, albeit at the expense of significantly larger computational overheads.

Minor modifications to the plane DDD simulation allow some aspects of dislocation-mediated deformation and residual stress generation to be captured. The so-called 2.5D model of Gaucherin at al. (2009) considers a simulation plane in which slip planes appear as line "traces," and dislocations are represented by the points at which they emerge in this plane (Fig. 7.12). The case of fcc crystal is considered, meaning that the 12 slip systems (defined as a distinct combination of slip plane normal and slip direction) appear as four line traces, the orientation of which depends on the simulation plane normal orientation with respect to the underlying crystal. Dislocation sources are represented by two dots, and their density can be varied, as can be the size of the simulation "box" that may represent a crystal with impenetrable or fixed boundaries. The use of this simulation approach allows revealing the propensity of dislocation arrays toward self-organization (patterning), and obtaining the dependence of yield and flow stress on grain orientation (Schmid factor) and grain size (Hall–Petch law). Furthermore, cyclic hardening and shakedown to elastic cyclic deformation can be observed at moderate plastic strain range.

As an accompanying result, DDD provides information about residual stress at the mesoscale. Because the simulation does not attempt to capture the actual dislocation structure, its interpretation requires statistical analysis or "coarsegraining" to take it up to the next structural level. This paradigm corresponds to the process modeling approach to residual stress analysis: to obtain the knowledge of residual stress distribution, the entire processing history must be known and simulated. The computing requirements for the implementation of this approach are not only very significant, but also require detailed knowledge of material parameters that affect its deformation response, which constitutes a major practical obstacle.

An alternative approach may be put forward that could be called "current state analysis" (as opposed to process modeling). The idea is not to rely on knowing or modeling the entire processing history, but barely to seek sufficient information in the form of a "snapshot" of the current residual stress state, representing the knowledge of some aspects at some selected points, and to form a satisfactory overall description of the residual stress on this basis.

There are two principal requirements crucial to implementing this agenda. First, residual stress evaluation needs to be conducted in the current state of the object, with an appropriate level of accuracy, sensitivity, and resolution. Second, the results obtained in this way must be processed by a suitable inverse model methodology to reconstruct the most likely overall residual stress distribution for the current state. These challenges are addressed in the following chapters.

## Reference

Cleveringa, H.H.M., Van der Giessen, E., Needleman, A., 2000. A discrete dislocation of analysis of mode I crack growth. Journal of the Mechanics and Physics of Solids 48, 1133–1157.

# CHAPTER 8

# Residual Stress "Measurement"

*J. Brodsky: Elegy for John Donne 1963*

*E. Chausson, Poème for Violin and Orchestra, op.25*
*Andrei Korsakov and USSR TV and Radio Large Symphony Orchestra (V. Fedoseyev) 1982*

*Mark Rothko, Black in deep red (oil on canvas, 276 × 136 cm) 1957*

As indicated in the Preface, putting the word "measurement" within quotation marks in the title of this chapter is indicative: residual stress, as indeed any stress, is not measureable directly, but must be deduced from measurement with the help of some suitable model. For this reason we will be talking of *residual stress evaluation*, to indicate that the interpretation of experimental results will be accompanied by a model, however simple, to correlate them with stress.

To make sense of the various approaches and scales of analysis we will classify the methods for stress evaluation according to scale, physical basis, resolution, sensitivity, precision, etc. The purpose of this classification will be to offer the reader a chance to appreciate the differences and complementarity between different approaches, and to equip them with the basis to make the most suitable selection for their studies.

## CONTENTS

## 8.1 CLASSIFICATION

Most residual stress evaluation methods actually rely on measuring *strain*, followed by the application of some deductive procedure to compute stress.

Strain can be calculated from the change in the distance between chosen points. Thus, it is possible to obtain strain by monitoring the distortion of a pattern deposited on the sample surface (e.g., a grid, a pattern of points, or a diffraction grating). This approach can yield excellent results for measuring deformation at the *surface*. However, this approach requires the knowledge of the reference state, and so cannot be used for *residual* stress.

A Teaching Essay on Residual Stresses and Eigenstrains. http://dx.doi.org/10.1016/B978-0-12-810990-8.00008-2

The methods that measure *strain increment* when residual stress is relieved by material removal belong to the family of *strain relief* methods. I prefer not to use the term "relaxation," because in other contexts in mechanics this is associated with time-dependent evolution of stress and strain, e.g., due to the phenomenon of creep. In contrast, "relief" refers to elastic unloading, and hence allows measurement of elastic strain increments. This reveals that the central idea of this approach is to observe how the strains sustained within the material due to the presence of residual stress are *relieved* by modifying the boundary conditions, notably, by introducing new surfaces.

Residual stresses are sustained inside the material as a result of the bonds present between different parts. When the bonds are removed, a new traction-free surface is created. If the deformation is monitored during this operation, then the previously existing stress state can be deduced.

In problem *A* illustrated in Fig. 8.1, residual stresses are present in the body. In *B* residual stresses have been partially relieved by cutting, resulting in the displacement $u'$ (total strain $\varepsilon'$, etc.). Displacement increment $(u' - u)$ and strain increment $(\varepsilon' - \varepsilon)$ can be monitored using some suitable means of measurement. Provided the relief is elastic, the preexisting stress can then be calculated with the help of a purely elastic model. The limitation to the method arises if unloading due to the creation of new surfaces causes inelastic deformation, e.g., due to reverse plastic flow. This introduces uncertainties that are usually deemed to be insurmountable, placing restrictions on the use of this approach.

This important point raises further considerations regarding the accuracy and reliability of strain relief methods. It is apparent from the description that residual stress *evaluation* is performed on the basis of certain explicit or implicit assumptions about the nature of the residual stress state before sectioning, and regarding the unloading process. These assumptions may be overly simplistic or restrictive, e.g., the shape of the cut may be idealized, introducing uncertainty in the interpretation. Often even the assessment of this uncertainty is problematic. A rational and consistent procedure for residual stress evaluation

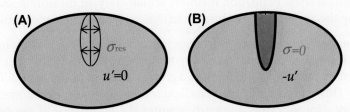

**FIGURE 8.1**
Illustration of residual stress evaluation by strain *relief*, which involves cutting (sectioning) and the measurement of strain increment.

by strain relief would make use of eigenstrain modeling of residual stress, in that it allows validating or refuting of the assumptions made.

Residual stress evaluation approaches that do not rely on monitoring the strain increment (relief) during material removal use coupling relationships between different physical properties of the object, on the one hand, and stress or strain, on the other, to obtain an estimate. Examples of such physical methods include magnetic techniques, ultrasonic techniques, Raman spectroscopy, and diffraction.

Residual stress is a quantity of great complexity, not only because of its inherent tensorial nature, but also because of the dependence on the scale of consideration. An important aspect of residual stress classification concerns the distinction made between residual stresses according to the length scale at which stress is perceived. In Chapter 2, the concept of stress was introduced with reference to an elementary volume of material by dividing the components of internal force acting across a selected cross-section by the corresponding area:

$$\sigma_{ij} = \frac{\Delta F_j}{\Delta A_i}. \tag{8.1}$$

This definition highlights the importance of structural scale of consideration involved in perceiving residual stress: choosing the size of the volume considered defines the scale of stress analysis. It is also important to note that the above definition implies *averaging*, i.e., any variation of stress over distances smaller than the length scale determined by the chosen cross-sectional area is deliberately ignored. Incidentally, a similar averaging operation is implicit in the definition of engineering strain and small-scale rotation:

$$\varepsilon_{ij} = \frac{1}{2}\left(\frac{\Delta u_i}{\Delta x_j} + \frac{\Delta u_j}{\Delta x_i}\right), \ W_{ij} = \frac{1}{2}\left(\frac{\Delta u_i}{\Delta x_j} - \frac{\Delta u_j}{\Delta x_i}\right). \tag{8.2}$$

It follows that the concepts of stress and strain are inherently scale dependent, and their interpretation and measurement must be carried out with explicit reference to the region over which they are being evaluated (the "gauge volume").

Fig. 8.2 introduces a popular classification of residual stresses according to length scales (probably originally due to E. Macherauch, and used widely by V. Hauk and L. Pintschovius). It is important to note that the stress separation into the three types is *additive*, namely:

$$\sigma^{RS} = \sigma^{RS,I} + \sigma^{RS,II} + \sigma^{RS,III}. \tag{8.3}$$

Thus, each further stress term refers to the deviation from the previous on a *finer scale*: Type II stresses indicate the extent to which the grain average value of residual stress differs from the macroscopic trend of Type I macrostress, whereas

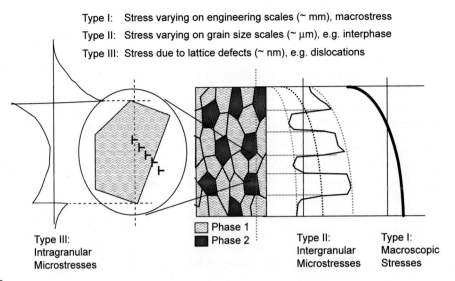

Type I:   Stress varying on engineering scales (~ mm), macrostress

Type II:  Stress varying on grain size scales (~ μm), e.g. interphase

Type III: Stress due to lattice defects (~ nm), e.g. dislocations

Phase 1
Phase 2

Type III:
Intragranular
Microstresses

Type II:
Intergranular
Microstresses

Type I:
Macroscopic
Stresses

**FIGURE 8.2**

Classification of residual stresses according to the length scale of consideration. *After Macherauch, E., 1987. Origin, measurement and evaluation of residual stresses. In: Macherauch, E., Hauk, V. (Eds.), Residual Stresses in Science and Technology, Proceedings of the International Conference on Residual Stresses, 1986 Garmisch-partenkirchen (FRO). DOM Informationsgesellschaft, Oberursel, p. 3.*

Type III gives the intragranular (down to nanoscale) deviation of stress from the grain average level. Type III may be directly linked to crystal defects and features such as vacancies, voids, precipitates, inclusions, slip bands, and dislocation pile-ups. Such direct association may not be possible for Type II and Type I stresses that arise from aggregation of finer scale sources (*strain nuclei*). Nevertheless, it is clear that these stresses may be associated with microscale inhomogeneities in the thermal and mechanical properties between phases or microstructural domains (grains) within the material (Type II), or with macroscopic regions of inelastic deformation arising from the deformation history, e.g., bending, pressing. The various sources of residual stress can be generalized based on the unifying, scale-independent concept of *eigenstrain*.

In summary, alongside macrostresses (Type I residual stress) perceived at the typical engineering length scales of consideration of millimeters, *microstresses* of Type II and Type III are defined. Type II microstresses relate to the scale of typical grain microstructure in engineering alloys, i.e., from hundreds and tens of micrometers down to single microns. Often Type II microstresses can be gauged by selective averaging offered, e.g., by beam methods utilizing spectroscopy or diffraction that possess sensitivity to material volumes with specific structure or orientation. Type III microstresses refer to intragranular length scales down to the order of nanometers.

It is appropriate to consider further refinement of the scale of consideration down to individual vacancies, dislocations, stacking faults, etc. A natural limit arises from the atomic structure of matter: once the characteristic dimension of the analysis becomes comparable with the interatomic distance (be it in crystalline or amorphous solid), the underlying "graininess" of matter commands that bonding forces are taken into consideration in place of "smeared" continuum description. An interesting set of challenges then arises in determining the extent to which continuum theories can capture the mechanical interactions at those shortest length scales. Seeking answers to these questions has been a growing theme in nanoscale materials science and technology, with both bottom-up and top-down approaches being pursued. Notable challenges in this field on which work is ongoing concern the interpretation of molecular dynamics simulations in terms of continuum stresses and strains. For our purposes it is worth mentioning that as long as the scale of consideration remains above several interatomic distances, i.e., a few nanometers, the possibility of interrogating internal forces and deformation using classical continuum definitions of stress and strain remains relevant.

In terms of the experimental techniques available for residual stress evaluation, both the effective gauge volume and the nature of measurement ought to be considered. According to the approach, residual stress evaluation techniques can be classified into:

1. *Nondestructive.* Physical analysis methods that allow residual stress evaluation via the quantification of changes in the structural or physical parameters, e.g., the interplanar lattice spacing by diffraction, or changes in molecular bond stiffness by spectroscopy. The wide range of experimental methods includes Raman spectroscopy and a multitude of diffraction techniques that make use of X-ray, neutron, or electron beams. Nondestructive methods rely on the quantification of the relative changes in the measured parameters and require comparison with a reference. The accuracy of these techniques is therefore limited by the precision to which reliable reference values can be determined.

2. *Destructive.* The introduction of traction-free surfaces by sample sectioning induces stress redistribution and strain relief in the surrounding material. This approach is therefore most closely related directly to the conventional definition of stress. The quantification of strain or displacement caused by the relief of tractions at the sectioned surface can be used in combination with numerical modeling to back-calculate the residual stresses originally present in the material. Experimental techniques that rely upon this approach include the slitting and contour method, with typical resolution in the range of fractions of a millimeter being achievable.

3. *Semidestructive.* The introduction of localized stress relief through hole drilling or core milling introduces minimal disturbance to the overall state but can be used to quantify locally the magnitude of residual stress. The approach is similar to destructive techniques, in that the quantification of strain change induced at the surface serves as an input for back-calculation of the residual stress. Conventional semidestructive techniques are capable of resolving stress at the resolution of $\sim 1$ mm laterally and with the depth resolution of $\sim 0.03$ mm. The need for probing residual stresses at smaller scales pushed researchers to develop procedures that involve finer tools for material removal and strain relief evaluation. Nanoscale Focused Ion Beam (FIB) ring-core milling in combination with Digital Image Correlation (DIC) allows quantifying residual stresses in precisely defined micron-sized gauge volumes. Comparison with simulation is required to relate the strain relief to the preexisting residual stress. The method known as FIB-DIC provides a conceptually appealing mechanistic approach that lends itself naturally to the application for the study of inter- and intragranular residual stresses (Type II and Type III).

From the various strain relief methods worth discussing in somewhat more detail we pick *layer removal* with curvature measurement, *hole drilling* and the *contour method*, noting that other techniques used are in many respects related or similar to these.

From the physical methods particular attention should be paid to *diffraction techniques* offering a unique nondestructive measurement ability combined with high precision and versatility, particularly in terms of sensitivity to distinct structural (crystallographic) phases, and spatial resolution.

Fig. 8.3 illustrates the range of scales of structural consideration introduced by Kassner et al. It is worth noting that the cartoons of structural organization perceived at progressively refined scales can be mapped directly to the above-mentioned definition of Type I macrostress, and Type II and Type III microstresses. With the introduction of this length scale–based classification, it is also possible to identify the key modeling and measurement approaches used at appropriate scales. For example, continuum level numerical modeling (macrostresses) uses the Finite Difference, Finite Element, or Discrete Element approach. Type II microstress analysis is often performed using Crystal Plasticity Finite Element approaches. Simulation of Type III microstress often involves discrete or continuum dislocation modeling, as well as Molecular Dynamics and ab initio calculations.

Owing to the specific aspects of implementation, not only the classification but also the discussion of experimental methods of strain and residual stress

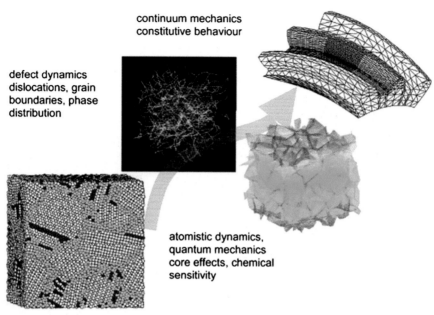

continuum mechanics
constitutive behaviour

defect dynamics
dislocations, grain
boundaries, phase
distribution

atomistic dynamics,
quantum mechanics
core effects, chemical
sensitivity

**FIGURE 8.3**

The range of scales for structural consideration of materials (Kassner et al., 2005).

evaluation should also be arranged according to the spatial resolution they afford. The present chapter is entirely devoted to the discussion of *macrostress* measurement. In view of the growing importance of microstress analysis, and the specific nature of experimental approaches used for this purpose, these matters are presented separately in Chapter 9.

## 8.2 LAYER REMOVAL AND CURVATURE MEASUREMENT

When layers are removed from one side of a flat plate containing residual stresses, the force balance is disturbed and must be recovered by plate bending. The curvature depends on the original stress distribution present in the layer that has been removed and on the elastic properties of the remainder of the plate. By carrying out a series of curvature measurements after successive layer removal, the distribution of stress in the original plate can be deduced.

For coating deposition, a variation of this technique is often used, whereby the curvature measurement of the substrate is monitored as successive layers of coating are applied. The method uses the same basic principles as the layer removal technique by measuring changes in curvature to deduce residual stress, but in this case these are due to coating deposition or growth.

The curvature of the specimen can be measured using a variety of methods including optical microscopy, laser scanning, strain gauges, or profilometry, depending on the resolution and range required, and the choice of instrument. Measurements are usually made on narrow strips to avoid multiaxial curvature and mechanical instability. This technique has been used successfully for polymeric composites, polymeric moldings, and ceramic coatings on metal substrates.

Consider a residually stressed plate of width $b$ and initial thickness $h$. If a layer of small thickness $dh$ containing residual stress $\sigma$ is removed, a small moment $dM$ will develop and result in curvature $\kappa$. We can find the rate of change of moment $M$ with thickness during layer removal and relate it to the residual stress $\sigma$. The relevant formulas already discussed in the context of elastic beam bending analysis are

$$M = \frac{EI}{R} = EI\kappa, \quad I = \frac{bh^3}{12} + \frac{bh^3}{4} = \frac{bh^3}{3}$$

$$M = \int_0^h \sigma z\, dz, \quad dM = \int_0^{h-dh} \sigma z\, dz \cong \sigma dh(h- dh/2) \cong \sigma h\, dh, \quad \frac{dM}{dh} = \sigma h$$

$$(8.4)$$

On the other hand,

$$\frac{dM}{dh} = E\kappa \frac{dI}{dh} + EI \frac{d\kappa}{dh} = Ebh^3 \left( \kappa + \frac{1}{3} \frac{d\kappa}{dh} \right). \qquad (8.5)$$

and therefore

$$\sigma = Ebh^2 \left( \kappa + \frac{1}{3} \frac{d\kappa}{dh} \right). \qquad (8.6)$$

Now if curvature $\kappa$ is measured continuously as a function of the plate thickness, then stress $\sigma$ can be determined using Eq. (8.6). Numerical differentiation is required to find the near-surface value. The method may be susceptible to errors from anticlastic bending, uneven material removal, etc.

## 8.3 HOLE DRILLING

Hole drilling is a widely used technique for measuring residual stress that is relatively simple, cheap, quick, and versatile. Equipment can be laboratory based or portable, and the technique can be applied to a wide range of materials and components. The principle of the technique involves the introduction of a small hole into a component containing residual stresses and subsequent measurement of the locally relieved surface strains. The residual stress can then be calculated from these strains using formulas and calculations derived from experimental measurements and/or Finite Element Analyses (Schajer, 1988).

Because any residual stress created by the drilling process will adversely affect the results, it is important to choose a suitable drilling method. This includes the use of conventional drilling, abrasive jet machining, and high-speed air turbines.

The hole drilling method has several limitations:

- In practical terms there is no point making measurements beyond a depth roughly equivalent to the drill diameter, because no additional strain can be measured.
- The basic hole drilling analysis assumes that the material is isotropic and linear elastic, that stresses do not vary significantly with depth, and that the variations of stress within the body of the hole are small.
- The basic analysis applies only where residual stress values do not exceed half the yield strength of the material.

*Deep hole drilling* is a variation of the technique, which has been developed for measuring residual stresses in thick-section components. The basic procedure involves drilling a small reference hole through the specimen and subsequent removal of a column of material, centered about the reference hole, using a trepanning technique. The diameter of the reference hole is measured accurately along its length before the column is machined out. When the column is removed, the stresses relax and the reference hole diameter and column dimensions change; the dimensions of the column and reference hole are then remeasured and the residual stresses calculated from the dimensional changes caused by removing the material from the bulk of the specimen. Using this technique the through thickness residual stress distribution can be measured on specimens up to 100-mm thickness.

## 8.4 THE CONTOUR METHOD

The principle of the contour method is as follows:

1. Consider a plate containing a residual stress distribution shown in Fig. 8.4.
2. Cut, with minimum disturbance, the part into two. Owing to the presence of residual stresses the newly created free surface will be distorted. A contour map of the surface height is made using profilometry.
3. Residual stress that was released during cutting corresponds to the distributed traction needed to bring the contour back to the flat. It can

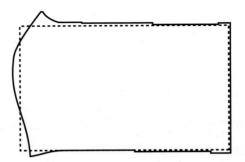

**FIGURE 8.4**

Illustration of the principle of the contour method. The shape before cutting is indicated by the *dashed line*.

be found by imposing the measured displacements with the negative sign within a Finite Element model.

The limitation of the contour method is that it is most naturally applied to bodies in which the residual stress state is independent on the length coordinate—a condition that may be described as *plane eigenstrain state*. The electric discharge machining sectioning approach also ignores displacements of material points parallel to the cut plane, and so would be unable to reconstruct, for example, residual stress states created in a shaft by inelastic torsion without deplanation. However, various modifications of this technique have been proposed, including successive sectioning by multiple planes that can allow the reconstruction of multiple stress components.

## 8.5   PHYSICAL METHODS

*Magnetic* properties have been utilized for residual stress measurements and their primary advantages are that they are nondestructive, cheap, simple, and rapid. Ferromagnetic properties of steels and other ferromagnetic materials are sensitive to the internal stress state because of magnetostriction and the consequent magnetoelastic effect. (Magnetostriction is the process whereby each magnetic domain is strained along its direction of magnetization).

At minimum energy the magnetization will align with the crystalline directions—the magnetic easy axes. A change in the stress level will result in a change in the number of domains aligned along each of the easy axes leading to a reduction in the magnetoelastic energy. Although the stress dependence of the magnetic parameters is quite strong, there are many other variables, such as hardness, texture, and grain size that also affect the measurement. For this reason, a combination of magnetic techniques is required so that the effect of these other variables can be eliminated. AEA Technology's MAPS System incorporates a combination of Stress Induced Magnetic Anisotropy together with Directional Effective Permeability methods into a single instrument.

*Ultrasonic* methods utilize the sensitivities of the velocity of ultrasound waves traveling through a solid to the stress levels within it. Changes in the speed of ultrasonic waves in a material are directly affected by the magnitude and direction of stresses present. Because the velocity changes are small and are sensitive to the material's texture (grain alignment) it is often more practical to measure transit times. Because the changes in velocity depend on the stress field over the entire ultrasonic path, the spatial resolution is poor. The acoustoelastic coefficients necessary for the analyses are usually calculated from calibration tests. Although ultrasonic methods provide a measure of the macro residual stresses over a large volume of material, the presence of texture in the material often restricts their spatial resolution. Nevertheless, they can be measured in the bulk of the material and are therefore well suited to routine inspection operations. In addition, the instrumentation is portable and quick to implement.

*Raman effect* involves the interaction of light with matter. Incident laser light causes the bonds between atoms to vibrate. Analysis of the scattered light, known as Raman spectrum, reveals vital information about a sample's physical state and chemical structure. This technique is nondestructive and noninvasive and has a high spatial resolution (down to a few microns). Raman or fluorescence lines shift linearly with variations in hydrostatic stress. This method has fine spatial resolution and by using optical microscopy it is possible to select regions of interest just a few microns in size. The method is essentially a surface strain measurement technique, but with optically transparent materials such as sapphire it is even possible to obtain subsurface information. Materials that give Raman spectra include silicon carbide and alumina-zirconia ceramics, and the method is particularly useful for studying fiber composites.

*Diffraction* has been a vast and crucially important theme in the modern physical and biological science, ever since the work of Röntgen, Laue, and Braggs established the possibility of interrogating atomic arrangements within solids without the ability to resolve individual particles. It is the development of this theme that enables perhaps the greatest discovery of all, the deciphering of the way DNA records the code for constructing living organisms.

Owing to the vast volume of literature available on X-ray diffraction and crystallography at large, and in application to residual stress analysis specifically, we will not devote space to these techniques here, except on occasions when we use them as a "gold standard" validation reference for other approaches. For the purposes of further discussion it suffices to say that scattering methods (including diffraction) allow precise determination of interatomic or interplanar distances, serving as a strain gauge for deformation analysis, provided its reference value is known somehow, this leads directly to the evaluation of elastic strain, and enables residual stress calculation.

**Table 8.1** Technique Selector (Kandil et al., 2001)

| Key Techniques | Size of Component | Contact or Noncontact | Destructive? | Laboratory Based or Portable | Availability of Equipment | Speed | Standards Available | Cost of Equipment | Cost of Measurement | Level of Expertise Required |
|---|---|---|---|---|---|---|---|---|---|---|
| | | | | | | **Practical Issues** | | | | |
| Hole drilling | Structure Artifacts Coatings | Contact | Semi | Both | Widespread | Fast/Medium | ASTM E837-99 | Low | £50–200 | Low/Medium |
| X-ray diffraction | Structures Artifacts coatings | Noncontact | No | Both | Generally available | Fast/Medium | No | Medium | £50–200 | Medium |
| Synchrotron | Artifacts Coatings | Noncontact | No | Laboratory | Specialist | Fast | No | Strategic/Government facility | High (exact range is not available) | High |
| Neutron diffraction | Artifacts | Noncontact | No | Laboratory | Specialist | Medium/Slow | No | Strategic/Government facility | £10–1500 | High |
| Curvature and Layer removal | Artifacts coatings | Contact | Yes | Laboratory | Generally available | Medium | No | Low | £50–200 | Low/Medium |
| Magnetic | Structures Artifacts | Noncontact | No | Both | Generally available | Rapid | No | Medium | Low (exact range is not available) | Low |
| Ultrasonic | Structures Artifacts | Contact | No | Both | Generally available | Rapid | No | Medium | Low (exact range is not available) | Medium |
| Raman/fluorescence | Structures Artifacts coatings | Noncontact | No | Both | Generally available | Fast | No | Medium | Low (exact range is not available) | Medium |

| Key Techniques | Resolution | Penetration | Stress Type | Stress State | Stress Gradient | Uncertainty | Sampling Area/Volume | Remarks |
|---|---|---|---|---|---|---|---|---|
| | | | | | **Physical Characteristics** | | | |
| Hole drilling | 50—100 μm depth increment | =Hole diameter | Macro | Uniaxial Biaxial | Yes, difficult to interpret | Varies with depth | 1–2 mm diameter. 1–2 mm deep | |
| X-ray diffraction | 20 μm depth 1 mm laterally | 5 μm, Ti 50 μm, A1 1 mm—layer remova1 | Macro micro | Uniaxial Biaxial | Yes, with layer removal | Limited by several factors | 0.1–1 mm² 0.05–0.1 mm | Resolution for steel 1 mm diameter (min) 15 μm depth |
| Synchrotron | 20 μm lateral to incident beam. 1 mm parallel | >500 μm 100 mm, Al | Macro micro | Uniaxial Biaxial Triaxial | Yes | Limited by grain sampling | 0.1 mm³ | Measures strain only. Rarely possible to obtain stresses |
| Neutron diffraction | 500 μm | 100 mm, Al 25 mm, Fe, 4 mm, Ti | Macro micro | Uniaxial Biaxial Triaxial | Yes | Limited by number of counts | >1 mm³ | |

| | | | | | | | |
|---|---|---|---|---|---|---|---|
| Curvature and layer removal | Not applicable | Depends on material and measurement method | | Macro | Uniaxial Biaxial | Yes | Layer removal can be combined with other measurement methods such as XRD and magnetic to give information on stress profiles |
| Magnetic | 1 mm | 20–300 μm (Barkhausen) | >2 mm$^2$ | Macro | Uniaxial Biaxial | No | Sensitive to microstructure, anisotropy, and texture |
| Ultrasonic | 5 mm | >100 mm along specimen | 1–400 mm$^3$ | Macro | Uniaxial Biaxial | No | |
| Raman/fluorescence | 0.5 μm | Surface | | Macro | Uniaxial Biaxial | | |

| Key Techniques | Material Type | Composites | Crystalline or Amorphous | Material Issues | | |
|---|---|---|---|---|---|---|
| | | | | Coated? | Surface Preparation | Surface Condition |
| Hole drilling | Metals Plastics Ceramics | Yes | Either | Yes | Light abrasion—strain gauge | Flat—preparation must not affect stresses |
| X-ray diffraction | Metals Ceramics | Yes | Crystalline | Yes | Important | Important |
| Synchrotron | Metals Ceramics | Yes | Crystalline | Yes | Important | Not critical |
| Neutron diffraction | Metals Ceramics | Yes | Crystalline | No | Not critical | Not critical |
| Curvature and layer removal | All | Yes | Not applicable | Yes | Not critical | Not critical |
| Magnetic | Ferromagnetic materials[1] | No | Crystalline | No | Not critical | Not critical |
| Ultrasonic | Metals Ceramics | Yes | Crystalline | Yes | Not critical | Not critical |
| Raman/fluorescence | Ceramics Polymers | Yes | Either | Yes | Not critical | Not critical |

| Technique | Pros | Cons |
|---|---|---|
| Hole drilling | • Quick, simple<br>• Widely available<br>• Portable<br>• Wide range of materials<br>• Deep hole drilling for thick section components | • Interpretation of data<br>• Destructive<br>• Limited strain sensitivity and resolution |
| X ray diffraction | • Versatile, widely available<br>• Wide range of materials<br>• Portable systems<br>• Macro and micro residual stresses (RS) | • Basic measure merits<br>• Laboratory-based systems, small components |
| Synchrotron | • Improved penetration & resolution of X-rays<br>• Depth profiling<br>• Fast<br>• Macro and micro RS | • Specialist facility only<br>• Laboratory based |

Continued

**Table 8.1** Technique Selector (Kandil et al., 2001) continued

| Technique | Pros | Cons |
|---|---|---|
| Neutron diffraction | • Excellent penetration & resolution<br>• 3D maps<br>• Macro and micro RS | • Specialist facility only<br>• Laboratory based |
| Curvature and layer removal | • Relatively simple<br>• Wide range of materials<br>• Can be combined with other techniques to give stress profile | • Limited to simple shapes<br>• Destructive<br>• Laboratory based |
| Magnetic | • Very fast<br>• Wide variety of magnetic techniques<br>• Portable | • Can apply to ferromagnetic materials only<br>• Need to separate the microstructure signal from that due to stress |
| Ultrasonic | • Generally available<br>• Very fast<br>• Low cost<br>• Portable | • Limited resolution<br>• Bulk measurements over whole volume |
| Raman/fluorescence | • High resolution<br>• Portable systems available | • Surface measurements<br>• Interpretation<br>• Calibration<br>• Limited range of materials |

## 8.6  METHOD OVERVIEW AND SELECTION

The macroscopic methods introduced previously, both strain relief and physical, tend to measure the average macroscopic (Type I) component of residual stress, leaving unanswered important questions about the residual stresses in *inhomogeneous* and *multiphase* materials.

Strain relief methods are more or less destructive by design. In terms of their accuracy they are approximately quantitative, because they rely on simplifying assumptions, such as uniform stress and moderate gradient.

Physical methods are nondestructive, but many of them are largely qualitative. For example, the acoustic and magnetic methods react strongly to the sample texture (preferred crystal orientation) and grain size, so calibration can be difficult, whereas spectroscopic methods tend to lack directional sensitivity.

*Diffraction* methods can overcome most of the limitations of the other techniques. Most significantly, alongside the ability to assess the macroscopic average elastic strains, they also allow *microscopic* residual strains to be evaluated.

Comparison between different methods is presented in Table 8.1, based on the publication from the UK National Physical Laboratory (NPL).

# CHAPTER 9

# Microscale Methods of Residual Stress Evaluation

*M. Lermontov, The hero of our time, 1841*

*Freddie Mercury/Montserrat Caballé, Guide me home/How can I go on, 1988*

*Arkhip I. Kuindji, Moonlit Night on the Dnepr (oil on canvas, 143 × 104 cm) 1882, Tretyakov Gallery, Moscow.*

## 9.1 PECULIARITIES OF RESIDUAL STRESS EVALUATION AT THE MICROSCALE

Localized stresses (at the micron scale or smaller) are often critical to the understanding of the origins and mechanisms of component failure. Unlike extrinsic properties such as stiffness, intrinsic properties of strength and fatigue resistance are dependent on the local "weakest link." This may be as small as a specific micron-sized region within a grain of material, at a grain boundary or junction. During service, it is the interaction between the residual stress and the applied load that determines the mechanical response and the likelihood of crack initiation. The interaction is played out over a wide range of different length scales, with the exact dominant dimension depending on the application. For example, it is customary to place principal emphasis on the macroscale behavior consideration in welding, micron scale in aeroengine assemblies, and submicron scale in the case of nanocomposites and electronic components. Residual stress evaluation at the microscale brings with it challenges that are specific and peculiar to structural analysis at this resolution. The progressive refinement of residual stress evaluation techniques has in the last few decades provided methods for the quantitative assessment of residual stress at resolutions down to a few microns and beyond. Some aspects of method development and application summarised in review articles (Lunt and Korsunsky, 2015; Lunt et al., 2015) are introduced below.

## 9.2 MICROFOCUS X-RAY DIFFRACTION METHODS

Collimation of a parallel X-ray beam to define the illuminated region of the sample can be effective down to the spot size of approximately $10 \times 10 \ \mu m^2$.

**109**

A Teaching Essay on Residual Stresses and Eigenstrains. http://dx.doi.org/10.1016/B978-0-12-810990-8.00009-4

Beam profiles smaller than this limit typically have insufficient flux to obtain diffraction profiles of the quality required for quantitative strain evaluation within a reasonable time period. Therefore, to reduce exposure times, tighter beam definition requires the use of focusing to improve the total X-ray flux into the gauge volume. This tight focusing can be accomplished using transmission optical elements, such as compound refractive lenses (CRLs), Fresnel zone plates, or reflective mirrors (Snigireva and Snigirev, 2006). There has been growing interest in the development and use of optical elements such as CRL-based transfocators (Vaughan et al., 2011) for high-energy X-rays, and the use of diamond for making kinoform lenses.

Monochromatic microbeam X-ray diffraction relies on obtaining a powder pattern from the gauge volume; therefore, the sample must have a fine-grained structure, with a submicron mean grain size. Debye–Scherrer patterns of coaxial scattering cones are typically collected on a two-dimensional (2D) detector mounted either in the direct beam (transmission, Fig. 9.1) or to a side (reflection). Procedures and software can be used to reduce these 2D diffraction patterns to the more conventional one-dimensional (ID) profiles (Davies, 2006), which in turn can be analyzed by full-profile refinement packages, such as FullProf, MAUD, or GSAS. These pattern refinements can be used to quantify the complete material strain state in the plane perpendicular to the incident beam, or as a measure of texture, i.e., orientation distribution function analysis. Nanocrystalline human dental tissues (dentine and enamel) offer a prominent example of a natural material in which synchrotron diffraction characterization has been applied to great effect.

In larger grained ($\sim$20–30 $\mu$m) microstructures, micron-sized beams can be used to evaluate intragranular strains and stresses. Using monochromatic beams for analysis of this kind presents a challenging need for precise rotational alignment between the crystallographic orientation and the incident beam, as well as micron-scale sample positioning. An alternative is offered by the use of white beam (polychromatic radiation), also known as Laue mode diffraction that can be carried out in reflection or transmission geometry. The analysis of 2D Laue diffraction images can be accomplished using automated software tools to determine grain orientation and deviatoric lattice strain. Pure hydrostatic expansion or contraction of a unit cell does not change the angles between lattice planes, meaning that the Laue pattern is not altered, although the energies corresponding to individual reflections are modified. This effect can be registered using an energy-resolving detector or by filtering the incident beam energy. Recent development of particular relevance to strain analysis concerns careful evaluation of error sources, and the quantification of small changes of Laue patterns due to lattice rotation and strain-induced distortion. Laue microdiffraction studies include the evaluation of stresses promoting the growth of tin whiskers, and multitechnique mapping of deformation of nickel polycrystals (Fig. 9.2).

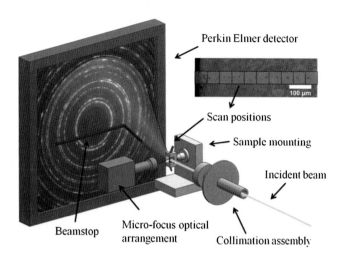

**FIGURE 9.1**

Schematic representation of the synchrotron X-ray powder diffraction setup at beamline I15 (Diamond Light Source, UK). X-ray diffraction patterns were collected by scanning the X-ray beam across the positions shown in the inset (Lunt et al., 2015).

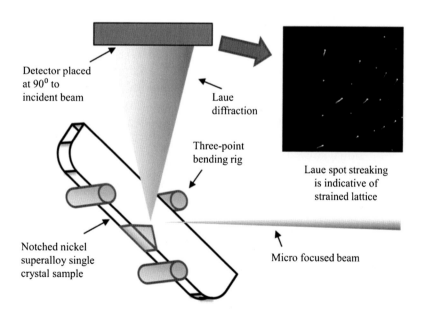

**FIGURE 9.2**

Schematic of Laue diffraction during in situ loading of a notched nickel superalloy single crystal (Lunt et al., 2015).

A note can be made regarding studies aimed at using atomic Pair Distribution Function (PDF) analysis to determine strain in noncrystalline materials. The paper by Poulsen et al. (2005) opened the way to the development of this approach, whereas a more recent publication by Huang et al. (2013) revealed the relationship between macroscopic strain and radial distribution peak shifts obtained from PDF analysis. Small Angle X-ray Scattering may also be used to evaluate nanoscale strains, although quantitative interpretation requires careful consideration.

## 9.3   ELECTRON DIFFRACTION METHODS

Electron diffraction provides another powerful route to determine lattice strains at resolutions ranging from the submicron scale in a back-scattered geometry to a few nanometers in transmission.

Electron Back-Scatter Diffraction (EBSD), originally discovered by Nishikawa and Kikuchi in 1928, has grown in popularity since the advent of commercial Scanning Electron Microscopes (SEMs) in the 1960s. Automatic pattern indexing using the Hough transform (Krieger Lassen et al., 1992) has since facilitated routine processing of large numbers of patterns. Grain orientation determination for microstructure mapping and microtexture analysis are now popular EBSD applications, along with the use of Kikuchi pattern quality assessment to visualize grain boundaries and plastically strained regions.

The classical angular resolution of EBSD lies in the range of $\sim 0.1°-0.5°$, and therefore the determination of lattice strain (typically required at a resolution $\sim 10^{-4}$) appears to represent a significant challenge using this technique. Nevertheless, particularly since year 2000, ever-increasing levels of EBSD sensitivity have been achieved through the improvement of interpretation procedures. Accurate intragranular lattice misorientation and quantitative residual stress analysis is now possible by using DIC to quantify small changes in Kikuchi patterns, the technique that became known as High Resolution EBSD (Wilkinson and Britton, 2012; Wilkinson et al., 2006) (Fig. 9.3).

Reports of high spatial-resolution ($\sim 10$ nm) (Keller and Geiss, 2012) EBSD mapping have been published using transmission geometry for thin samples such as transmission electron microscopy (TEM) lamella (Suzuki, 2013). This approach is similar to the high-resolution strain mapping performed using TEM transmission diffraction, e.g., using convergent beam diffraction (Vigouroux et al., 2014). In both these cases, care must be taken to account for stress relaxation during lamella preparation (Clément et al., 2004). Other notable methods utilizing scattering include convergent beam electron diffraction and the interpretation of high-resolution TEM images using Fast Fourier transform.

**(A)**

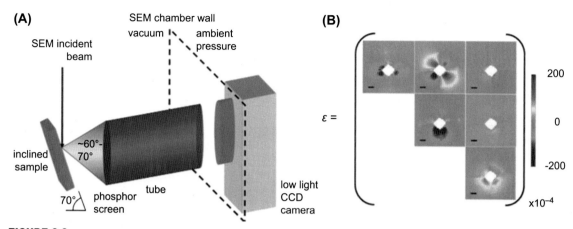

**(B)**

**FIGURE 9.3**

(A) Schematic diagram showing the experimental setup for Electron Back-Scatter Diffraction (EBSD). (B) EBSD elastic strain measurements around an indent in a silicon single crystal (Wilkinson and Britton, 2012). *SEM*, scanning electron microscope.

## 9.4 SPECTROSCOPIC METHODS

Spectroscopic techniques probe the atomic or molecular energetic characteristics of the sample to extract indirect information about residual stress state within the gauge volume. For example, peak shifts in Raman spectra can be related to the residual stress state within the volume illuminated by a monochromatic laser and confocal optics allow this beam to be focused to ~0.2 μm. Furthermore, due to the penetration of light through the surface material layers, such setups facilitate residual stress depth profiling of micron-sized material volumes (Roberts et al., 2014). Raman spectroscopy residual stress mapping has been used to study zirconia-based thermal barrier coatings (Liu et al., 2014) and the impact of etching on porous silicon (Kang et al., 2005) (Fig. 9.4). An important development in this field is correlative Raman and SEM imaging achieved by combining the two microscopy modes within one instrument (Correlative Raman-SEM microscopy (RISE microscopy) in life sciences, 2015).

All the various modes of micron-scale stress evaluation described earlier, although benefiting from the nondestructive nature, suffer from a number of limitations. First, not all material types can be studied using these techniques: for example, diffraction works only with crystalline samples that contain grains of a size that can be approximated as either a powder or as a single crystal (relative to the beam size). Furthermore, diffraction of heavily deformed materials (e.g., metallic alloys) that contain significant lattice distortion results in degraded diffraction patterns, making them impossible to interpret.

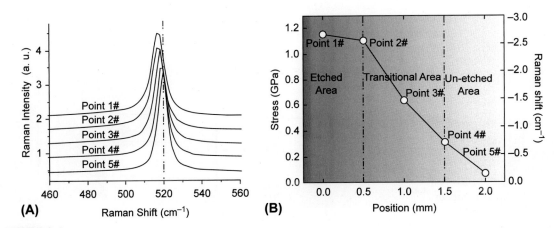

**FIGURE 9.4**

Spatially resolved residual stress analysis in an etched surface of porous silicon (Kang et al., 2005). The Raman peak shift detected at each location is shown in (A), and the corresponding residual stresses are shown in (B).

Spectroscopic methods such as Raman can be applied only to materials that contain molecular bonds. Consequently, polymers and oxides can be analyzed in this way, whereas metallic alloys cannot.

Most importantly, all beam-based methods provide *relative* stress measurement, i.e., require a reference state for reliable interpretation. Providing reliable micron-sized reference volumes is known to be formidably difficult in many samples of interest.

In contrast, techniques for stress evaluation based on material removal (e.g., slitting, sectioning, drilling) do not suffer from the above-mentioned limitations: they can be applied to both amorphous and crystalline materials, including after heavy plastic deformation. Furthermore, experimental studies (Baimpas et al., 2013) and numerical simulation (Korsunsky et al., 2009) have demonstrated the micro-ring-core method provides "on-board" built-in reference. This approach not only provides a reference for quantitative interpretation of diffraction and spectroscopy data, but also allows the determination of *absolute* residual stress state.

## 9.5 INTRODUCTION TO FIB-DIC MICROSCALE RESIDUAL STRESS ANALYSIS AND ERROR ESTIMATION

The idea of obtaining a minimally destructive probe of local residual stress using a ring-core geometry is not new: it goes back to the pioneering work of Keil in the early 1990's (Keil, 1992). The surge of recent interest is based on the work by

Korsunsky et al. (Korsunsky et al., 2009), in which microscale Focused Ion Beam (FIB) ring-core milling was used to quantify residual stress in a precisely defined gauge volume. This broadly applicable basis has facilitated the development of a range of similar related techniques (Winiarski et al., 2012a). These methods rely on the introduction of new traction-free surfaces with quantification of the resulting surface strain relief. SEM images of the surface are typically postprocessed using Digital Image Correlation (DIC) analysis software to quantify such changes. Comparison with Finite Element (FE) simulations is then used to relate this strain relief measurement to the preexisting residual stress value.

One of the main limitations on the resolution of these semidestructive techniques is the residual stress locally induced by ion implantation (Wang et al., 2006). The magnitude and region of this influence is dependent upon the milling geometry, ion energy, and material of interest (Oka et al., 2004). Typically this zone lies in the range 10–100 nm and therefore this acts as a limit to FIB-based stress quantification.

Microscale versions of the traditional macroscopic semidestructive analysis techniques have been shown to be effective for quantifying fine-scale residual stress, e.g., using slot milling (Prime, 1999) and hole drilling (Schajer, 1988). FIB-based microslitting was first published by Kang et al. in 2003 and has since become widely used to estimate residual stress in a direction perpendicular to the slit (Sabaté et al., 2006a,b, 2007a). Improvements have provided estimates of residual stress variation with depth (Winiarski et al., 2012a) and along the length of the slit (Mansilla et al., 2015). Microscale FIB hole drilling, on the other hand, has provided estimates of the 2D stress state in a number of applications (Vogel et al., 2006), including incremental depth-resolved analysis (Winiarski and Withers, 2010) and spatially resolved investigations (Winiarski et al., 2010). The main limitation of these approaches is that they rely on the strain relief induced in relatively large surface regions (typically tens of microns or more). This means that the exact region of the stress evaluation is often difficult to pinpoint, and the spatial resolution is consequently reduced. In turn, this somewhat limits their applicability for high-resolution spatially resolved analysis, i.e., marker interaction is guaranteed at very small length scales ($<10$ μm).

Novel FIB approaches based on the uplift or in-plane relief of surface material have been proposed to quantify the plane stress state; these include the micro-cantilever (McCarthy et al., 2000) and H bar (Krottenthaler et al., 2013) methods. The relatively long sample preparation times and limited measurement positions associated with these techniques means that they have limited relevance for stress mapping.

The ring-core FIB-DIC methodology allows quantification of the complete in-plane residual stress state at the micro- to nanoscale (Korsunsky et al., 2009,

2010; Song et al., 2011, 2012b; Sebastiani et al., 2011; Lunt and Korsunsky, 2014). This technique relies on inducing and measuring the strain relief within a well-defined gauge volume: a micropillar that is FIB milled in the material surface down to a depth-to-diameter aspect ratio ~1. The high speed, ease of application, precision, and nanometer placement accuracy of this approach has meant that it has since been applied to a wide range of materials and problems.

The use of a small island defines a precise gauge volume, but limits the area over which DIC can be performed. Despite this restriction, repeated imaging of the island surface provides a more thorough record of the strain change as a function of milling depth. The strain relief profile is then fitted with a "master curve" based on the results of previous FE simulations of the milling process (Gelfi et al., 2014). This approach means that FE analysis is not required, and quantitative results are obtained in minimal time.

Advances in this technique have demonstrated that depth-wise spatially resolved analysis can be implemented using the FIB-DIC approach (Salvati et al., 2014a). Although the theoretical framework for this depth profiling is well established and robust, experimental practicalities such as ion implantation, surface roughness and degradation, and material inhomogeneity (Martins et al., 2010) impact the results of this approach. Therefore, attempts to declare residual stress depth profile measurements must be treated critically and by validation against the limited number of comparable techniques.

Alongside the review of the available microscale residual stress measurement techniques, the present chapter assesses the robustness and validity of the FIB-DIC methods to quantify in-plane spatially resolved residual stress through the ring-core approach. Two possibilities arise for attaining this objective, namely, *sequential milling* or *parallel milling* of features on the sample surface. A schematic of these two approaches is shown in Fig. 9.5. *Sequential* milling involves the incremental determination of residual stress in islands placed at regular intervals (left of Fig. 9.5), whereas *parallel* milling requires simultaneous milling of multiple features in a contiguous array (right of Fig. 9.5).

The underlying principle of FIB milling for residual stress evaluation is the introduction of microscale traction-free surfaces at the location of interest. A range of experimental techniques has been proposed, differing in terms of the FIB milling geometries designed to measure a particular aspect of the stress state.

The introduction of traction-free surfaces results in the reequilibration of the stress state in the region of interest. This can be observed as a strain change in the regions neighboring the milling location and can be recorded using high-resolution SEM imaging.

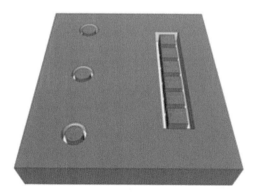

**FIGURE 9.5**

Schematic of sequential (left) and parallel (right) Focused Ion Beam—Digital Image Correlation milling approaches Lunt and Korsunsky (2015).

To quantify the resulting strain relief, DIC analysis is performed on the sequence of SEM images of the sample surface. Comparisons between the displacement fields observed and the results of analytical or FE models are drawn, typically using optimization to fit a parametric description of the strain relief observed. Back-calculation of the residual stresses originally present in the surface can then be performed using these relief estimates.

One of the key requirements of microscale residual stress analysis is the precise determination of the gauge volume position and size. This requirement is very important in spatially resolved residual stress analysis, which has previously been shown to be critical in understanding failure in a wide range of samples (Lunt et al., 2015). The nanoscale precision associated with FIB milling and SEM imaging means that the gauge volume can be quantified to submicron accuracy and therefore techniques based on these technologies have the capabilities of providing insight at the scale required.

In the past few decades, micron-scale residual stress analysis has increasingly become accessible via X-ray diffraction and infrared/visible light spectroscopy methods. These techniques have potentially large gauge volumes through-thickness, i.e., in the direction of the beam and are further restricted in two main respects. First, many rely upon the interaction between the incident radiation and crystal lattices. Therefore, noncrystalline materials or materials that experienced severe deformation and associated lattice distortion cannot be analyzed using these techniques. The second limitation is associated with the need for reliable estimates of unstrained reference samples, or lattice constants. Although relative variations in lattice parameter can be determined without this insight, it is not possible to carry out *absolute* residual stress determination, which is the quantity of most interest in terms of component failure and lifetime estimation.

In contrast with the aforementioned limitations, the FIB milling and DIC analysis provides a means of absolute residual stress evaluation (Lunt and Korsunsky, 2014). This approach is also not limited to crystalline materials, and can be applied equally well to amorphous or polymeric substrates (Baimpas et al., 2014b).

Apart from possessing demonstrable tangible benefits in terms of experimental capability, FIB milling and DIC techniques are also highly practical: the technical improvements and cost reduction of SEM and FIB microscopy systems have resulted in an increased access to high-specification multibeam systems necessary to perform FIB milling and DIC analysis. This increased access, in combination with the relatively short experiment durations (typically 10–50 milling locations can be interrogated in a single day) has promoted technique development and wider adoption. Importantly, the reduction in the experimental milling time is reflected in the vastly reduced nominal cost per stress measurement, making the approach highly competitive, especially when compared with synchrotron or other large facility–based techniques.

A key aspect of the FIB milling and DIC techniques is their interpretative nature, giving rise to the need for careful error analysis and uncertainty quantification, because the final use of the results in design practice relies crucially on the reliability, repeatability, and accuracy of stress evaluation. Therefore, attention needs to be given to the identification and quantification of generic uncertainties that arise in the context of FIB-DIC techniques (Lunt et al., 2015; Winiarski and Withers, 2015), with particular emphasis on the error estimation and propagation involved in strain quantification and stress evaluation. The fundamentals are reviewed, and case study examples of microscale FIB-DIC ring-core analysis are provided in the following discussion.

The FIB milling and DIC-based techniques are subject to a number of sources of experimental error that can influence the results obtained. These errors can be grouped into three main types: (1) errors associated with sample preparation, surface patterning, and the FIB milling process, (2) errors associated with the DIC analysis, and (3) errors associated with the residual stress quantification from the surface relief recorded. Although many techniques have been developed to moderate these errors, little has been published on this topic. Therefore, this important aspect has been selected as the main focus of this review.

The error of FIB milling and DIC-based analysis induced either by sample preparation or the experimental process typically has greater magnitude than that associated with postprocessing of the data. However, a significant number of publications have been devoted to reducing this error through the optimization of different geometries (Kang et al., 2003; Sabaté et al., 2007a; Korsunsky et al., 2009; Krottenthaler et al., 2013).

SEM imaging and FIB milling are both reliant upon the acceleration and focusing of electrons and ions. This results in relatively strong electromagnetic interaction with the sample surface. In the case of a highly insulating sample, or one in which a clear conduction path to the earth connection is not present, the buildup of charge can influence the path of incident electrons used for imaging and thereby cause the apparent image distortion and sample drift. As previously noted in the literature, this effect reduces the quality of SEM images (Okai and Sohda, 2012). Furthermore, the effect on the ion beam may manifest itself as a drift in the milling position. Overall, these effects reduce the effectiveness of DIC marker tracking. Several techniques have been used to reduce the impact of charging, such as improved mounting and the introduction of conduction paths by sputtering a thin (a few nanometers) conductive coating (for example, Au or Pd), or applying conductive paint to provide conductive path close to the region of interest (Suzuki, 2002), and charge neutralizer systems can also be used (Moncrieff et al., 1978). Variable pressure SEM systems can be used to improve imaging, but cannot be operated during FIB milling, thereby limiting the applicability of this technique (Newbury, 2002). Drift correction by tracking of a fiducial marker has also been shown to be successful for small amounts of drift (Marturi et al., 2013).

Another effect widely known to influence the SEM imaging quality of the DIC tracking regions is the redeposition of material sputtered by milling and not fully removed by the vacuum system. The deposition process reduces the marker tracking stability and thereby increases the error of the final stress estimate. Several approaches can be used to minimize the effect of surface quality degradation: FIB milling at reduced milling rates (i.e., low ion beam current) results in reduced redeposition at the cost of increased milling duration (Bhavsar et al., 2012). Refinement or reduction of the amount of milled material also reduces material redeposition on the sample surface. An alternative approach pioneered by Sebastiani et al. involves applying a deposition barrier between the DIC region and the milling location (Sebastiani et al., 2011). This approach has proved to be successful at reducing redeposition; however, it increases the experiment time and may influence the residual stress state before measurement.

The process of ion beam milling involves bombarding the surface of the sample with high-energy ions to effect material removal. Although on the macroscale this approach is very gentle, locally this ion interaction causes ion implantation accompanied by a cascade of material damage that is likely to induce residual stress in the sample. The thickness of the layer affected by ion damage can be up to $\sim$30 nm in materials such as silicon, depending on the ion energy, target material, and angle of incidence. This phenomenon introduces a lower bound on FIB-DIC milled features of $\sim$0.5 µm, below which ion milling itself begins to affect the residual stress state perceived. The impact and location of the ion radiation damaged zone are also dependent upon the choice of milling geometry, with some techniques proving to be more robust than others.

The FIB milling and DIC residual stress analysis technique is intrinsically a surface analysis technique. However, in many cases the location of the most critical residual stress lies below the surface. Therefore, to access this location, sample sectioning has to be performed. This sectioning process relieves the stresses in the direction of the normal of the newly created surface, thereby modifying the residual stress state in the area of interest. Simple approximations can be developed based on the relationship between the initial and modified stress state using the comparison between plane strain and plane stress relationships, respectively (Lunt et al., 2016). However, to understand this interaction fully, FE simulations are required involving intrinsic parameters such as eigenstrain (Song et al., 2012a; Korsunsky et al., 2007).

The consequence of sample sectioning needs to be carefully assessed in terms of the residual stress induced during sample preparation. Gentle removal techniques are typically used to remove material from the sample surface, for example, electrical discharge machining (Jameson, 2001), diamond sawing (Bravman and Sinclair, 1984), or grinding. Incremental fine grain polishing, heavy ion beam polishing (Grogan et al., 1992), or electropolishing (Landolt, 1987) can then be performed to reduce the depth of the residually stressed region. The exact magnitude of the compressive stress induced and the region over which this influence extends is highly dependent upon the substrate material and polishing regimen. However, FIB-DIC techniques are highly sensitive to near-surface effects, and therefore this behavior is known to influence the measure of residual stress originally present in the sample.

One final difficulty that has previously been identified in the use of FIB milling and DIC-based techniques is the limitation on sample surface topology (Winiarski and Withers, 2015). Minor roughness (on the order of a few tens of nanometers) can be accommodated (Lunt and Korsunsky, 2014), e.g., by adjusting the reference position of the surface in the course of interpretation. However, residual stress analysis on fracture surfaces and other highly uneven samples is likely to require careful approaches. Surface polishing can be used to reduce surface roughness at the expense of time and the potential modification of the preexisting residual stress.

## 9.6    FIB-DIC MILLING GEOMETRIES

Advances in FIB software and hardware have in recent years greatly expanded the range of geometries that can be milled using ion beams. This flexibility has enabled a wide range of FIB milling and DIC residual stress analysis techniques to be developed. Each technique has been designed to meet the needs of the specific application or to quantify a particular stress state of interest, and therefore has particular advantages and limitations when compared with others.

## 9.6.1  Surface Slotting

FIB microsurface slotting was proposed by Kang et al. in 2003 and is based on the miniaturization of the macroscale crack compliance method (Prime, 1999). The approach is based on milling a single narrow rectangular slot into the material surface (Fig. 9.6A). To quantify the displacement field, DIC is performed on SEM images of the regions on either side of the slot. A comparison between the relief observed and FE methods is then used to back-calculate the average residual stress in the direction perpendicular to the long axis of the slot. The simple milling regimen and speed of analysis are the main benefits associated with this technique, and developments have extended this approach to provide

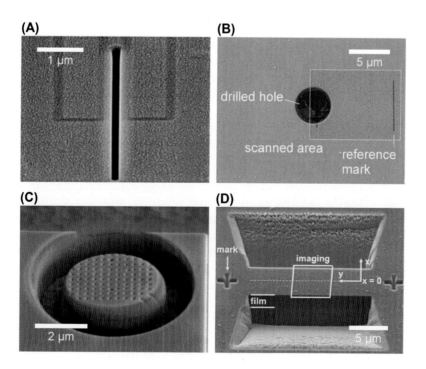

**FIGURE 9.6**

Focused Ion Beam—Digital Image Correlation residual stress analysis geometries. (A) Microslotting geometry (Kang et al., 2003). (B) Microhole drilling (Sabaté et al., 2007a). (C) The ring-core approach (Korsunsky et al., 2009). (D) H-bar geometry (Krottenthaler et al., 2013). *From Lunt, A.J.G., Korsunsky, A.M., 2015. A review of micro-scale focused ion beam milling and digital image correlation analysis for residual stress evaluation and error estimation. Surf. Coatings Technol. 283, 373–388. http://dx.doi.org/10.1016/j.surfcoat.2015.10.049.*

depth-resolved (Winiarski et al., 2012a) and spatially resolved analyses (Winiarski et al., 2010; Mansilla et al., 2015).

In addition to the obvious limitation of providing only a single component measure of residual stress, microslotting is known to be limited in a number of other ways. First, this technique relies on performing DIC on a region outside the "nominal" gauge volume. Residual stresses in this surrounding region are known to influence the relief observed and therefore this greatly reduces the precision of the gauge volume definition both in terms of size and position. The geometry also greatly limits how close subsequent markers can be placed, and therefore restricts the use of this technique for highly spatially resolved analysis. Another limitation of the approach is the need for repeated FE simulations (Winiarski et al., 2012a; Mansilla et al., 2015), which increases processing time, complexity, and cost.

### 9.6.2 Hole Drilling

FIB microhole drilling was originally published by Sabaté et al. in 2007a and is based on the miniaturization of the classical macroscale blind hole drilling method originally developed by G. Schajer (1988). This approach provides a measure of the 2D stress state in the surface of interest by milling a small circular hole into the sample surface (Fig. 9.6B). The DIC of the surrounding region is used to quantify the strain relief field induced by milling to provide an estimate of the stress state originally present. Microscale hole drilling has the benefits of being experimentally simple and fast, and depth-resolved extensions of this technique have been reported (Winiarski and Withers, 2012).

Despite these improvements, microscale hole drilling relies on the DIC analysis of the stressed regions outside the gauge volume location, and therefore is subject to reduced gauge volume precision in a way similar to microslotting. Another difficulty associated with this experimental approach is the limited amounts of strain relief induced by milling. The impact of noise on the perceived residual stress value(s) is therefore far greater than for other methods, and the solution of the inverse problem associated with this technique requires careful regularization (Winiarski and Withers, 2012). Regularization not only reduces unrealistic variations in the stress values obtained, but also limits the sensitivity of the technique.

### 9.6.3 Microscale Ring-Core Milling

The microscale ring-core FIB milling and DIC approach was originally proposed by Korsunsky et al. in 2009 and is based on the miniaturization of the macroscopic ring-core method originally developed by Keil (1992). In this experimental approach, an annular milling pattern is used to obtain full strain relief of a core of material lying at the sample surface (Fig. 9.6C).

This approach ensures that the gauge volume (central core) is well defined both in terms of position and size. Owing to the isotropic milling geometry, by examining the strain relief in the core in different orientations the ring-core technique can provide a quantitative measure of the complete 2D in-plane stress state in the sample of interest. Other advantages of this geometry include the fact that the strain relief observed in the core is typically much larger in magnitude than that for the microscale hole drilling (exterior field) and microslotting methods. Because the displacement values are larger, the method is more robust with respect to noise. It has also been shown that the strain relief induced across the center of the core is predominantly uniform (Salvati et al., 2014a), ensuring that effective averaging can be used to improve the robustness of interpretation. Finally, the use of a central core, rather than a reliance upon the surrounding region means that markers can be placed close together for high-resolution spatially resolved analysis (Lunt et al., 2015). Depth-resolved analysis has also been demonstrated using this technique (Bemporad et al., 2014).

There is a direct link between the main advantage and the principal difficulty associated with the microscale ring-core FIB milling geometry, namely, that the region over which DIC analysis is performed is typically smaller compared with that in other techniques. To overcome this limitation, an approach based on incremental milling and repeated SEM imaging has been developed (Song et al., 2011, 2012b; Korsunsky et al., 2010). This methodology provides a record of the average strain relief within the micropillar island (or core) as a function of milling depth which is then compared with the results of FE modeling. The use of multiple milling steps and images greatly improves the precision with which strain can be determined, in contrast with the simple comparison between "before milling" and "after milling" SEM images. Further improvement is afforded by analyzing subsequent images, thereby improving the reliability of DIC marker tracking.

### 9.6.4 H-bar Milling

The H-bar FIB milling and DIC geometry was originally proposed by Krottenthaler et al. in 2012 (Krottenthaler et al., 2013). This approach is based on milling two trenches on either side of the gauge volume (Fig. 9.6D) using the automated TEM lamella preparation methods typically found in FIB-SEM systems. The estimate of the residual stress component in the direction perpendicular to the long axis of the trench is then obtained. In comparison with the microscale slotting method, the H-bar has a more precise gauge volume definition both in terms of position and size. Similar to the ring-core methodology, this approach also benefits from uniform strain relief across the central bar, and from the increased magnitude of the surface strain relief. This geometry also enables DIC analysis to be performed over relatively large regions, thereby improving DIC marker averaging compared with the ring-core approach.

Although the main limitation of the H-bar approach is that the technique provides stress quantification only in 1D, a publication by Sebastiani et al. extended this technique to 2D by milling further lines to leave a fully relieved core in the form of a square (Sebastiani et al., 2014). This final geometry is very similar to the ring-core approach. Note that the two-stage milling process (first parallel bars, then additional cross lines) also provides insight into Poisson's ratio at the microscale.

## 9.7    ERROR ESTIMATION AND PROPAGATION IN FIB-DIC RESIDUAL STRESS ANALYSIS

The quantification and propagation of DIC marker fitting standard deviation value(s) is a necessary step required to obtain consistent confidence bounds for residual stress (Kang et al., 2003; Mansilla et al., 2015). A typical result of this type of analysis is shown in Fig. 9.7, showing error values between 14% and 36% found in interpreting the data, with the average value of 22%. Limited use of outlier removal in this analysis may account for a large scatter and relatively high magnitude of experimental error.

The introduction of improved error propagation and assessment facilitates technique development and improvement. Often these techniques require

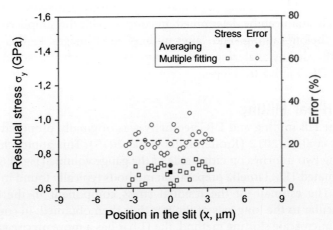

**FIGURE 9.7**

Spatially resolved stress analysis using microslitting Focused Ion Beam—Digital Image Correlation (DIC) in steel. Errors are based on the percentage difference between a least-squares fit and results at each position, giving the average magnitude of 22%. DIC marker standard deviation was taken into account in this analysis, but complete error propagation was not performed (Mansilla et al., 2015).

regularization, as shown in the case of depth-resolved microscale hole drilling in Fig. 9.8 (Winiarski and Withers, 2012). The 90% confidence intervals for this technique increase from 25% of the declared value in the near surface region, up to around 200% of the value at full depth. These figures demonstrate the high levels of sensitivity of this technique and highlight the importance of precise error quantification in FIB milling and DIC residual stress analysis presented in the following discussion (Baimpas et al., 2014b).

DIC image processing is used to determine the relative displacement field between two or more images. This technique is well established and the fundamental principles (Hild and Roux, 2012; McCormick and Lord, 2010), coding implementations (Blaber et al., 2015; Eberl et al., 2006), and potential applications (Chu et al., 1985; Sutton et al., 2009) have been discussed at length.

The displacement field can be decomposed into the effects of image shift, distortion, rotation, or magnification. Analysis is typically performed using one of two approaches: full-field DIC (Pan et al., 2007; Wang and Cuitiño, 2002) involves applying an unknown displacement field to the entire original image to estimate the average global change or subset DIC (Pan et al., 2008; Yaofeng and Pang, 2007), which is associated with placing markers ("seeds") onto the original image, each of which is associated with a particular correlation subset window. The displacement field of each of these subset windows

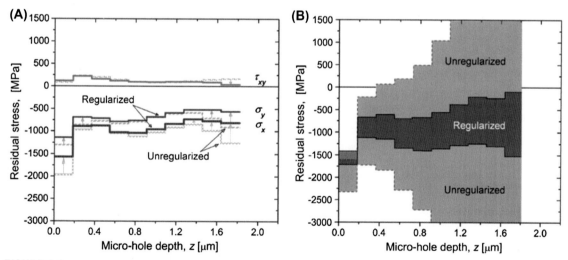

**FIGURE 9.8**

Depth-resolved residual stress analysis using the microhole Focused Ion Beam milling and Digital Image Correlation (DIC) technique in a bulk metallic glass. (A) The full 2D in-plane stress state has been incrementally quantified and regularization has been used to overcome the intrinsic sensitivity of this technique. (B) The 90% confidence intervals for stress in the $x$ direction are indicated and correspond to 25%–200% of the value obtained. DIC marker standard deviation was taken into account in this analysis and complete error propagation was performed (Winiarski and Withers, 2012).

can typically be determined in a computationally efficient and fast manner compared with full-field DIC. The flexibility associated with subset DIC, in terms of marker placement, size, and overlap, along with the ability to assess and eliminate poorly tracked markers, has meant that this DIC approach has become increasingly robust, reliable, and efficient.

DIC is typically based on the approximation that the displacement field can be represented by a vector shift between the two data sets. For this approximation to be valid, image distortion and rotation need to be small over the subset window size. This approach enables precise and rapid estimation of the relative displacement through the use of 2D normalized cross-correlation $C(u, v)$ (Chu et al., 1985). In the case of subset window half-width of $n$ pixels (total size of $(2n + 1) \times (2n + 1)$ pixels of the subset window $\mathscr{S}$) and an image $\mathscr{J}$ at the relative offset of $u, v$ is given by:

$$C(u,v) = \frac{\sum_{i=-n}^{n} \sum_{j=-n}^{n} \left[\mathscr{S}(i,j) - \overline{\mathscr{S}}\right] \cdot \left[\mathscr{J}(u+i, v+j) - \overline{\mathscr{J}}_{uv}\right]}{D(u,v)}, \tag{9.1}$$

where

$$D(u,v) = \sqrt{\sum_{i=-n}^{n} \sum_{j=-n}^{n} \left[\mathscr{S}(i,j) - \overline{\mathscr{S}}\right]^2 \cdot \sum_{i=-n}^{n} \sum_{j=-n}^{n} \left[\mathscr{J}(u+i, v+j) - \overline{\mathscr{J}}_{uv}\right]^2}, \tag{9.2}$$

and $\overline{\mathscr{S}}$ is the mean intensity value of the subset, and $\overline{\mathscr{J}}_{uv}$ is the mean intensity value within a $(2n + 1) \times (2n + 1)$ window within $\mathscr{J}$ that is centered at $u, v$. This expression can be repeatedly applied across the region to determine the values of the cross-correlation function over the given range of $u$ and $v$ displacements, as shown in Fig. 9.9. The peak position on this surface corresponds to the $u, v$ combination that achieves the most similarity between the subset windows chosen within the reference and deformed images. The magnitude of the peak is the correlation coefficient, $r$. It is a measure of similarity between the subset windows in the reference and deformed image, with the value of unity indicating that the two subset windows are identical.

Postprocessing of the shifts determined at each marker location can be used to quantify global displacement fields between the initial and subsequent images. Typically in FIB residual stress analysis, the DIC strain resolution sought is $10^{-4}$ or better; this corresponds to 1 pixel displacement over 10,000 pixels. This high precision has previously served as the main limitation in the use of DIC to quantify strain relaxation. Despite this, the combination of four approaches now enables this resolution to be obtained routinely:

1. *Subpixel resolution DIC.* The vector shift is calculated using integer pixel displacement values of $u$ and $v$. To improve the precision of shift determination, the estimation of the peak position can be obtained

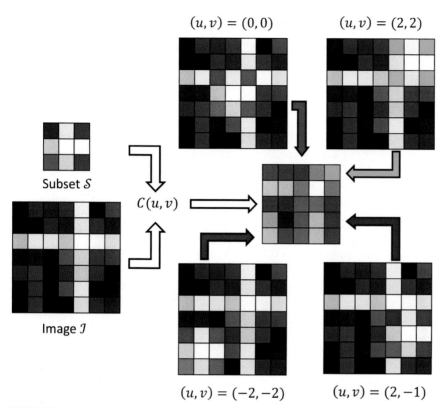

$(u, v) = (0, 0)$        $(u, v) = (2, 2)$

Subset $\mathcal{S}$

$C(u, v)$

Image $\mathcal{J}$

$(u, v) = (-2, -2)$        $(u, v) = (2, -1)$

**FIGURE 9.9**

Schematic of normalized 2D cross-correlation, $(u, v)$, between subset $\mathcal{S}$ and image $\mathcal{J}$. Each pixel value on the cross-correlation surface is associated with a different relative offset $(u)$ between the $\mathcal{S}$ and $\mathcal{J}$ as shown on the right side of the figure (Lunt and Korsunsky, 2015).

by functional fitting of the cross-correlation surface, e.g., using a polynomial fit to describe the region closest to the peak. In addition to providing a more statistically rigorous estimate (through the use of multiple data points), this approach can be used to determine the peak center to subpixel resolution, typically to better than 0.05 pixels. Such approaches also facilitate the estimation of confidence bounds for the relative displacements between images.

2. *DIC marker averaging.* The displacement or strain relief induced during FIB milling of different geometries have been fully characterized either through analytical or FE techniques. Typically these relief profiles have been described as functional representations, and therefore averaging can be performed on the marker displacements obtained. For example, in the microslotting method the displacement at a given distance from the slot edge is expected to be constant and

therefore an average value along the length of the slot can be obtained. Least-squares fitting of the linear relationship between these displacement measurements and the results of FE analysis can then be used to provide the final estimate of strain relief (Winiarski and Withers, 2012). The average displacement of many hundreds of markers can therefore be used to further improve the precision of the DIC strain relief.

3. *High-magnification SEM imaging.* To optimize the accuracy of converting displacements determined by DIC into strain, high-magnification and high-pixel matrix SEM imaging is necessary. Typically, DIC analysis is performed on SEM images of the size greater than $1024 \times 1024$ pixels, with the nominal pixel size of 5 nm or less.

4. *High imaging contrast.* To optimize 2D image correlation, high-contrast imaging is necessary. This ensures that the cross-correlation surface peak is sharp, so that subpixel peak displacement measurement is valid. High-contrast imaging may be possible because of the inherent sample nature, but surface patterning is typically used on homogenous substrates. This can take the form of the application of nanoparticles or highly textured coatings, or can be based on FIB or electron milling or deposition of features (Winiarski et al., 2012b). However, care must be taken to minimize the impact of surface treatment on the residual stress state measured, and to minimize the possibility of aliasing in DIC analysis.

Precise quantification of error bounds is critical for providing meaningful residual stress estimates using any analytical technique, from macroscale hole drilling (Schajer and Altus, 1996) to powder diffraction—based techniques (Noyan and Cohen, 2013). This requirement is equally true in microscale FIB milling and DIC-based methods, and the analysis of error estimation and propagation that needs to be conducted is similar for many of these approaches.

## 9.7.1    DIC Error Estimation

Accurate residual stress error quantification in FIB milling and DIC techniques depends upon the quality of the SEM images collected during the experimental process. The precision of the resulting DIC analysis is highly sensitive to the image noise, beam focus, and contrast, and therefore error estimation must start from this stage of the analysis. As stated previously, the most widely used correlation approach is subset DIC. Hence this section focuses solely on the errors associated with the normalized 2D cross-correlation approach.

When sharp, high-contrast, low-noise images are captured, the cross-correlation surface typically takes the form of a single, sharp, and well-defined peak as shown in Fig. 9.10A. The peak correlation coefficient in this

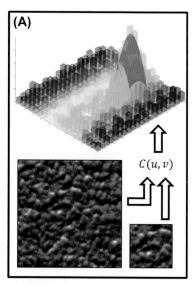
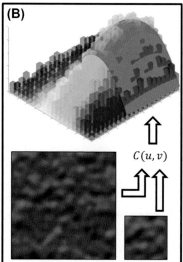
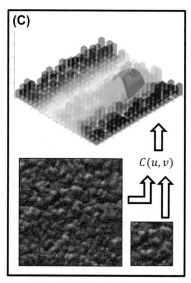

**FIGURE 9.10**

The normalized cross-correlation surface (bar plots) and polynomial peak fitting (green (gray in print versions) surfaces) associated with three different image types (Lunt and Korsunsky, 2015). (A) Sharp, low-noise images result in a single well-defined peak with a correlation coefficient ~1 and precise estimates of the relative offset between images. (B) Poorly focused, blurry, or low-contrast images result in a broad peak with a correlation coefficient ~1 but imprecise estimates of the relative offset between images. (C) Noisy images result in a low correlation coefficient and a high roughness cross-correlation surface, which lead to imprecise estimates of the relative offset between images.

case is close to unity, and the sharpness of the peak provides a precise estimate of the relative offset.

In the case of a poorly focused, blurred, or low-contrast image, significant peak broadening is observed in the cross-correlation surface as illustrated in Fig. 9.10B. Even though the correlation coefficient in this case may also be close to unity, the precision of the displacement evaluation between the two images is much reduced.

Cross-correlation of images containing large amounts of noise will result in a high-roughness cross-correlation surface as shown in Fig. 9.10C. Noise also produces random differences between the reference and deformed image subsets, thereby reducing the magnitude of the correlation coefficient at the peak. These two effects lead to a large decrease in the signal to noise ratio, and a reduction in the precision of displacement determination between the two images.

Examination of the three image cases outlined in Fig. 9.10 demonstrates that the correlation coefficient on its own does not provide a good measure of the precision of displacement determination. Despite this, this parameter is a useful thresholding measure to remove noisy data points.

In addition to providing subpixel resolution DIC, peak fitting also gives an estimate of the standard deviation of peak position. This is a much more reliable measure of displacement determination precision, and therefore is critical for error estimation and the removal of poorly tracked markers.

The large number of markers used in DIC analysis means that the computational cost associated with tracking each marker needs to be carefully considered. Depending upon the DIC precision required, the complexity of the fitting process can be tailored using two main approaches. First, a subset of the cross-correlation surface can be selected for analysis, rather than fitting the entire surface. Choosing a region centered on the peak position can vastly reduce implementation time. Second, the 2D function selected for fitting can be chosen depending on the precision required. For example, a 2D Gaussian function typically provides a good match to the peaks observed in the cross-correlation surface; however, the computational cost associated with fitting this function is higher than a polynomial peak fit.

In the case of the microscale ring-core FIB-DIC approach, the DIC software developed by Eberl et al. (Eberl et al., 2006) is based on 2D polynomial peak fitting of a $3 \times 3$ window around the highest magnitude point in the cross-correlation surface. Least-squares fitting is performed using matrix multiplication to determine the profile of the peak in the form

$$C_P(u,v) = A_1 + A_2 u + A_3 v + A_4 uv + A_5 u^2 + A_6 v^2 \tag{9.3}$$

where $A_i$ for $i = 1{:}6$ are the variables to be determined, $u$ is the horizontal displacement and $v$ is the vertical displacement. Differentiation of this function gives the expressions for the peak position:

$$u_P = \frac{2A_6 A_2 - A_3 A_4}{A_4^2 - 4A_5 A_6}, \quad v_P = \frac{2A_5 A_3 - A_4 A_2}{A_4^2 - 4A_5 A_6}. \tag{9.4}$$

In this coding implementation, rounding is then used to determine this figure to a resolution of 1/1000th of a pixel. An estimate of the covariance matrix can also be obtained using further matrix multiplication and the residuals at each data point (Farebrother, 1988). The diagonal terms in this matrix represent the variance $\sigma_{Ai}^2$ of each of the variables $A_i$ for $i = 1{:}6$. Examination of the off-diagonal terms reveals that the five variables ($A_i$ for $i = 2{:}6$) used in the calculation of $u_P$ and $v_P$ can typically be approximated as statistically independent. The propagation of these errors can then be used to estimate $\sigma_{up}$ and $\sigma_{vp}$, the standard deviations of $u_P$ and $v_P$, respectively:

$$\sigma_{up} = \sqrt{u_P^2 \left[ \frac{4\left(A_2^2 \sigma_{A6}^2 + A_6^2 \sigma_{A2}^2\right)^2 + \left(A_4^2 \sigma_{A3}^2 + A_3^2 \sigma_{A4}^2\right)^2}{\left(2A_6 A_2 - A_3 A_4\right)^2} + \frac{16\left(A_5^2 \sigma_{A6}^2 + A_6^2 \sigma_{A5}^2\right)^2 + 2A_4^4 \sigma_{A4}^4}{\left(A_4^2 - 4A_5 A_6\right)^2} \right]},$$
$$\tag{9.5}$$

$$\sigma_{vp} = \sqrt{v_p^2 \left[ \frac{4\left(A_5^2\sigma_{A3}^2 + A_3^2\sigma_{A5}^2\right)^2 + \left(A_4^2\sigma_{A2}^2 + A_2^2\sigma_{A4}^2\right)^2}{\left(2A_5A_3 - A_4A_2\right)^2} + \frac{16\left(A_5^2\sigma_{A6}^2 + A_6^2\sigma_{A5}^2\right)^2 + 2A_4^4\sigma_{A4}^4}{\left(A_4^2 - 4A_5A_6\right)^2} \right]}$$

(9.6)

Although several approximations are necessary in the use of the above-mentioned expressions, the simplicity of these calculations ensures that the peak center position and associated standard deviations can be computed with a minor increase in processing time. These estimates of confidence can then be used in the removal of outliers.

## 9.7.2 DIC Outlier Removal

To improve the accuracy and precision of residual stress estimates, the removal of poorly tracked DIC markers (or outliers) is a necessary step in the reliable estimation of the resultant displacement field. Ineffective tracking can be caused by a variety of different effects, including aliasing, redeposition of FIB milled material onto the sample surface, noisy imaging, or changes to the surface appearance near milling locations (Huang et al., 1997). A wide range of different techniques, each with their own strengths and limitations, can be used to remove outliers. Experience demonstrates that a combination of these methods typically provides the best possible overall result.

1. *Correlation coefficient thresholding.* The maximum correlation coefficient, $r$, can be used as a threshold to remove markers. A typical threshold for the correlation coefficient is 0.5, which can be used as a simple technique to leave only markers with a high-magnitude peak in the cross-correlation surface.

2. *Peak position standard deviation thresholding.* The standard deviations of the estimates of peak position ($\sigma_{up}$ and $\sigma_{vp}$) provide estimates of the sharpness of the peak in the cross-correlation surface. Relatively high standard deviations in either the $u$ or $v$ direction indicate a poorly tracked marker, and therefore thresholding can be used to leave only those markers with high precision peak estimation. A typical histogram of peak position standard deviation for the ring-core FIB milling and DIC approach is shown in Fig. 9.11.

3. *Markers moving relative to neighbors.* The vector displacement fields ($u_P$, $v_P$) can be postprocessed to identify markers that appear to undergo a large displacement relative to their neighbors, and thereby remove those erroneous points. A normalized dot product between each vector and the average shift of several of its neighbors has been widely used.

4. *Outliers to expected displacement fields.* FE modeling or analytical calculation has previously been used to characterize the displacement fields associated with each type of FIB milling and DIC geometries.

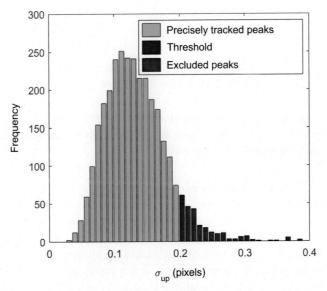

**FIGURE 9.11**

A typical histogram of $\sigma_{up}$, the peak center position obtained during the Digital Image Correlation (DIC) analysis of the ring-core Focused Ion Beam milling and DIC approach (Lunt and Korsunsky, 2015). Thresholding has been performed at the average value of $\sigma_{up}$ to leave only the precisely tracked markers.

Typically, least-squares fitting of the expected displacement field profile is used to determine one or more scaling parameter values. This fitting process can be used to estimate the distance between each marker and the expected profile. Thresholding can then be used to remove markers that are far from the nominal distribution. This process can be effective in removing any remaining outliers but care must be taken to ensure that the initial fit is representative of the behavior observed. This process must therefore be performed only after other outlier removal techniques have been applied.

In the case of the ring-core FIB-DIC technique, uniform strain assumption is often used, so that the expected displacement field is linear across the central core. The distance between markers and this linear trend can be used to remove outliers as shown in Fig. 9.12C. A value equal to 1.5 × average distance offers a robust threshold.

5. *Manual marker removal.* Manual examination of the strain or displacement fields produced by DIC serves as an effective approach in revealing outliers and poorly tracked markers. Typically, a peak in the strain or displacement field is indicative of a poorly mapped region or marker and can therefore be manually removed from the data set. These regions are often associated with aliasing, the impact of FIB redeposition or the surface profile changes within the milled region.

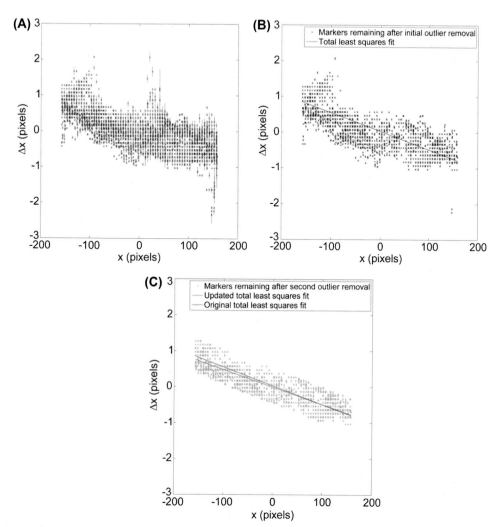

**FIGURE 9.12**

Outlier removal and total least-squares fitting used in ring-core Focused Ion Beam—Digital Image Correlation technique. Plots of displacement $\Delta x$ against position $x$ are shown along with 95% confidence intervals on each data point (Lunt and Korsunsky, 2015). (A) Markers remaining after correlation coefficient thresholding at 0.5. (B) Markers remaining after $\sigma_{up}$ and $\sigma_{vp}$ thresholding, along with the removal of markers moving relative to each other. Manual marker removal is also performed at this stage along with a total least-squares fit of the linear profile expected. (C) Markers remaining after thresholding of the distance from the original least-squares fit. The fit gradient change is seen to be small.

### 9.7.3    Strain Error Estimation

The FIB-DIC technique uses repeated SEM imaging of the core to obtain a record of relaxation at different milling depths. Strain relief observed at the core surface has previously been shown to be approximately uniform (Salvati et al., 2014a), corresponding to a linear variation in the displacement of DIC markers with position in the central region (Fig. 9.12) (Lunt et al., 2015). It illustrates the three-stage least-squares fitting and outlier removal regimen. Outlier removal of all DIC markers is initially performed using thresholding of the correlation coefficient, $r$ (Fig. 9.12A). Peak position standard deviation thresholding and the removal of markers moving relative to neighbors are then performed (Fig. 9.12B). Least-squares fitting is then performed on the remaining markers, and thresholding of the distance between each marker and this fit is used to leave only those markers that match well the form of the expected displacement field profile (Fig. 9.12C). The gradient of the final total least-squares fit provides estimates for the magnitude ($\Delta\varepsilon$) and standard deviation ($\sigma_{\Delta\varepsilon}$) of the strain relief for every selected orientation and image number.

### 9.7.4    Relief Profile Fitting

The fundamental basis of all FIB-DIC residual stress analysis techniques is the relationship between the stress state originally present within the gauge volume and the observed strain relaxation. This relationship is based on accurate quantification of the displacement field. For example, in the case of microslotting, the gradient of the linear displacement profile expected in the region surrounding the slot is matched directly to the results of FE simulations.

To improve the robustness of strain relief estimation, incremental milling can be performed. In FIB-DIC ring-core milling, the strain relief at each step ($\Delta\varepsilon$) is used to provide a record of relaxation as a function of milling depth. The general form of this strain relief profile (so-called master curve) was originally determined through FE simulations of the milling process (Korsunsky et al., 2009), and a functional representation of this profile has been proposed (Korsunsky et al., 2010):

$$f(\Delta\varepsilon_\infty, z) = 1.12\Delta\varepsilon_\infty \times \frac{z}{1+z}\left[1+\frac{2}{(1+z^2)}\right], \tag{9.7}$$

where $z = h/0.42d$, $h$ is the milled depth, $d$ is the core diameter, and $\Delta\varepsilon_\infty$ is the full strain relief at an infinite milling depth. For large milling depths, the value of the relief curve tends toward a plateau at $\Delta\varepsilon_\infty$ as shown in Fig. 9.13. The full strain relief parameter $\Delta\varepsilon_\infty$ forms the basis of the average stress calculation as described later.

Typically only nominal estimates of the milling depth are known and therefore one further parameter ($\eta$) is often introduced to relate the milling depth to the

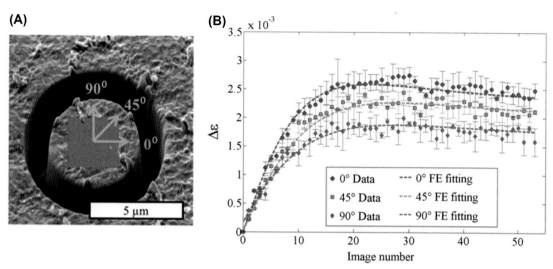

**FIGURE 9.13**

(A) A scanning electron microscopy image showing typical marker placement on the core center (red (gray in print versions)), along with the directions of strain relief quantification highlighted (Lunt et al., 2015). (B) Strain relief curves obtained as a function of milling depth in the 0°, 45°, and 90° orientations. The error bars indicate the 95% confidence intervals of each data point and the fitted profiles are indicated by the *dashed lines* (Lunt et al., 2015). *FE*, finite element.

image number ($I$), such that $h = \eta I$. Setting this parameter to a constant implies that the milling rate remains constant throughout the process. Although this is approximately true in the near-surface region, increasing deviation from this assumption may be observed at large depths.

To quantify the estimates of $\Delta \varepsilon_\infty$ in a given direction, nonlinear weighted least-squares fitting is performed of the master curve function to experimental data. The weighting parameters for this analysis are given by the inverse of the strain relief standard deviation ($\sigma_{\Delta \varepsilon}$) at each milling depth. The covariance matrix produced provides estimate of the standard deviation of $\Delta \varepsilon_\infty$ ($\sigma_{\Delta \varepsilon_\infty}$).

A key advantage of the ring-core FIB-DIC approach is that it provides full quantification of the 2D in-plane stress tensor, allowing identification of principal orientations and magnitudes of full strain relief at infinite milling depth ($\theta$, $\Delta \varepsilon_\infty^1$ and $\Delta \varepsilon_\infty^2$, respectively). To quantify these parameters, full strain relief analysis must be performed for at least three different directions. Typically, these can be chosen at an angular offset of 0°, 45°, and 90° to the horizontal direction (as shown in Fig. 9.13) to determine full strain relief parameters $\Delta \varepsilon_\infty^{0°}$, $\Delta \varepsilon_\infty^{45°}$, and $\Delta \varepsilon_\infty^{90°}$.

To propagate error estimates for each of the three directions, $\sigma_{up}$ and $\sigma_{vp}$ should be used in the calculation of the strain relief curves and associated confidence

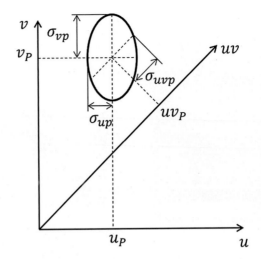

**FIGURE 9.14**

Schematic representation of the calculation of $u_\theta$ (denoted here by $uv_P$) and the associated uncertainty $\sigma_{uvp}$, the peak shift and standard deviation in the corresponding direction (Lunt and Korsunsky, 2015). The *ellipse* is representative of the uncertainty field in peak position in a given direction.

intervals in the 0° and 90° directions, respectively. Peak shift ($u_\theta$) for arbitrary orientation $\theta$ is given by

$$u_\theta = u_P \cos \theta + v_p \sin \theta. \tag{9.8}$$

An estimate of the standard deviation in any direction can be obtained by considering an elliptical representation of the uncertainty field in which the principal axes are aligned with the $u$ and $v$ directions (Fig. 9.14). For example, the uncertainty for 45° direction can be calculated as:

$$\sigma_{45°} = \sqrt{\frac{2\sigma_{up}^2 \sigma_{vp}^2}{\sigma_{up}^2 + \sigma_{vp}^2}}. \tag{9.9}$$

The DIC value and standard deviation in these three orientations must then be used to estimate the strain relief ($\Delta\varepsilon$) and standard deviation ($\sigma_{\Delta\varepsilon}$) in each direction and for every image. The resulting strain relief curves provide estimates of full strain relief for each orientation ($\Delta\varepsilon_\infty^{0°}$, $\Delta\varepsilon_\infty^{45°}$, and $\Delta\varepsilon_\infty^{90°}$), with the associated standard deviations of these measures $\left( \sigma_{\Delta\varepsilon_\infty^{0°}}, \sigma_{\Delta\varepsilon_\infty^{45°}}, \text{ and } \sigma_{\Delta\varepsilon_\infty^{90°}} \right)$.

The principal strain relief values and orientation can be evaluated using Mohr's circle for strain. The resulting expressions are

$$\Delta\varepsilon_\infty^{1,2} = \frac{\Delta\varepsilon_\infty^{0°} + \Delta\varepsilon_\infty^{90°}}{2} \pm \frac{1}{\sqrt{2}} \sqrt{\left(\Delta\varepsilon_\infty^{0°} - \Delta\varepsilon_\infty^{45°}\right)^2 + \left(\Delta\varepsilon_\infty^{45°} - \Delta\varepsilon_\infty^{90°}\right)^2}, \tag{9.10}$$

$$\theta = \frac{1}{2} \tan^{-1} \left( \frac{\Delta\varepsilon_\infty^{0°} - 2\Delta\varepsilon_\infty^{45°} + \Delta\varepsilon_\infty^{90°}}{\Delta\varepsilon_\infty^{0°} - \Delta\varepsilon_\infty^{90°}} \right). \tag{9.11}$$

By making the approximation that each term is statistically independent, estimates of the standard deviations of $\Delta\varepsilon_\infty^{1,2}$ and $\theta$ $\left( \sigma_{\Delta\varepsilon_\infty^{1,2}} \text{ and } \sigma_\theta, \text{ respectively} \right)$ can be obtained:

$$\sigma_{\Delta\varepsilon_\infty^{1,2}} = \frac{1}{2} \sqrt{ \sigma_{\Delta\varepsilon_\infty^{0°}}^2 + \sigma_{\Delta\varepsilon_\infty^{90°}}^2 + \frac{\left(\Delta\varepsilon_\infty^{0°} - \Delta\varepsilon_\infty^{45°}\right)^2 \left(\sigma_{\Delta\varepsilon_\infty^{0°}}^2 + \sigma_{\Delta\varepsilon_\infty^{45°}}^2\right) + \left(\Delta\varepsilon_\infty^{45°} - \Delta\varepsilon_\infty^{90°}\right)^2 \left(\sigma_{\Delta\varepsilon_\infty^{45°}}^2 + \sigma_{\Delta\varepsilon_\infty^{90°}}^2\right)}{\left(\Delta\varepsilon_\infty^{0°} - \Delta\varepsilon_\infty^{45°}\right)^2 + \left(\Delta\varepsilon_\infty^{45°} - \Delta\varepsilon_\infty^{90°}\right)^2} }, \tag{9.12}$$

$$\sigma_\theta = \frac{1}{\sqrt{2}} \frac{\sqrt{\sigma_{\Delta\varepsilon_\infty^{0°}}^2 \left(\Delta\varepsilon_\infty^{45°} - \Delta\varepsilon_\infty^{90°}\right)^2 + \sigma_{\Delta\varepsilon_\infty^{45°}}^2 \left(\Delta\varepsilon_\infty^{0°} - \Delta\varepsilon_\infty^{90°}\right)^2 + \sigma_{\Delta\varepsilon_\infty^{90°}}^2 \left(\Delta\varepsilon_\infty^{0°} - \Delta\varepsilon_\infty^{45°}\right)^2}}{\left(\Delta\varepsilon_\infty^{0°} - \Delta\varepsilon_\infty^{45°}\right)^2 + \left(\Delta\varepsilon_\infty^{45°} - \Delta\varepsilon_\infty^{90°}\right)^2}. \tag{9.13}$$

## 9.7.5 Calculation of Residual Stress

The relationship between surface strain relief and the residual stresses originally present in the surface forms the fundamental basis of FIB milling and DIC experimental techniques. In all techniques, this relationship is based on the elastic stiffness constants of the substrate material, which follows from the assumption that no reverse plastic flow occurs during FIB milling. In the case of microscale ring-core FIB-DIC geometry, this assumption is well justified, because material removal causes proportional unloading toward the origin (unstressed state). To calculate reliable estimates of the confidence bounds on residual stress, the uncertainties of both surface relief estimate and the elastic stiffnesses must be taken into account.

There are two principal origins of elastic modulus uncertainty. The first is the underlying uncertainties in material composition, processing history, and characterization. Elemental doping (Wachtel and Lubomirsky, 2011), microstructure (Suresh et al., 2015), thermal processing history (Nava, 1998), manufacturing route (Yu, 2009), and thermodynamic phase (Chaves et al., 2015) are all known to have a large impact on the elastic behavior. For example, elastic modulus of steel can vary in the range 190−220 GPa depending on the elemental doping and thermal history. For this reason an accurate record of processing history or precise materials characterization is necessary to reduce the uncertainty of description of the elastic response.

The other origin of elastic stiffness uncertainty is relevant to the study of anisotropic materials, in which stiffness is dependent upon the orientation. Elastic anisotropy can be the result of underlying crystallinity (Antunes et al., 2008), preferred grain orientation (or texture) (Qiu et al., 2011), or composite structure (Mortazavian and Fatemi, 2015). For example, directional stiffness in zirconia ceramic is known to vary by a factor of 2.6 depending upon the orientation in which load is applied (Lunt et al., 2014).

Various techniques have been developed to quantify the elastic behavior at the microscale, providing insight into the mechanical behavior at the length scale probed by FIB-DIC. Young's modulus evaluation at the microscale can be put into two distinct experimental groups. The first is based on the use of microscale testing devices to apply precise loads and to measure the resulting micro- to nanoscale displacements, for example, using miniaturization of the bulk scale tensile loading test (Sharpe et al., 1997; Read et al., 2001), microbeam bending or deflection (Hoffmann and Birringer, 1995; Tomioka and Yuki, 2004), nanoindentation (Korsunsky and Constantinescu, 2006, 2009; Tricoteaux et al., 2010; Antunes et al., 2007; Jennett et al., 2004), atomic force microscopy (Cho et al., 2005), the so-called thin-film biaxial bulge test (Vlassak and Nix, 1992; Kalkman et al., 2003), and micropillar compression (Lunt and Korsunsky, 2015; Mohanty et al., 2014). Because the geometry required for micropillar compression is identical to the annular arrangement in the ring-core FIB-DIC approach, the integration of these two techniques has the potential to provide the measures of the strain relief, contact modulus, and Poisson's ratio within a single experimental session. The second approach to Young's modulus determination is based on the interaction between deformation waves and the sample. For example, surface acoustic waves (Wang and Rokhlin, 2002; Flannery et al., 2001; Shan et al., 2011) and resonant ultrasound spectroscopy (Liang and Prorok, 2007; Pestka et al., 2008) have been used to quantify Young's modulus at the microscale.

Poisson's ratio is another parameter required for the conversion from surface relief to residual stress. Different approaches used to quantify this parameter at the microscale use nanoindentation (Jennett et al., 2004), atomic force microscopy (Cho et al., 2005), thin-film "bulge test" (Vlassak and Nix, 1992; Kalkman et al., 2003), surface acoustic waves (Shan et al., 2011), X-ray diffraction (Chang et al., 2009; Renault et al., 1998), bidirectional thermal expansion (Ye et al., 2006) or FIB milling (Sebastiani et al., 2014).

Assuming homogeneity and isotropy at the length scales under consideration, expressions for the nonequibiaxial principal stresses in the surface ($\sigma_1$ and $\sigma_2$) based on the ring-core FIB-DIC strain relief are (Korsunsky et al., 2010)

$$\sigma_1 = -\frac{E}{(1-v^2)}\left[\Delta\varepsilon_\infty^1 + v\Delta\varepsilon_\infty^2\right], \quad \sigma_1 = -\frac{E}{(1-v^2)}\left[\Delta\varepsilon_\infty^2 + v\Delta\varepsilon_\infty^1\right], \quad (9.14)$$

where $E$ and $v$ are Young's modulus and Poisson's ratio, respectively. By making the approximation that these terms are statistically independent, estimates for the standard deviation of the principal stresses ($\sigma_{\sigma1}$ and $\sigma_{\sigma2}$) are obtained from the standard deviation of Young's modulus ($\sigma_E$), Poisson's ratio ($\sigma_\nu$), and strain relief values from DIC $\left(\sigma_{\Delta\varepsilon_\infty^{1,2}}\right)$:

$$\sigma_{\sigma_1} = \frac{E}{(1-\nu^2)} \sqrt{\left[\left(\frac{\sigma_E}{E}\right)^2 + 2\left(\frac{\nu\sigma_\nu}{1-\nu^2}\right)^2\right](\Delta\varepsilon_\infty^1 + \nu\Delta\varepsilon_\infty^2)^2 + \sigma_{\Delta\varepsilon_\infty^1}^2 + \left(\nu\sigma_{\Delta\varepsilon_\infty^2}\right)^2 + \left(\Delta\varepsilon_\infty^2 \sigma_\nu\right)^2},$$

(9.15)

$$\sigma_{\sigma_2} = \frac{E}{(1-\nu^2)} \sqrt{\left[\left(\frac{\sigma_E}{E}\right)^2 + 2\left(\frac{\nu\sigma_\nu}{1-\nu^2}\right)^2\right](\Delta\varepsilon_\infty^2 + \nu\Delta\varepsilon_\infty^1)^2 + \sigma_{\Delta\varepsilon_\infty^2}^2 + \left(\nu\sigma_{\Delta\varepsilon_\infty^1}\right)^2 + \left(\Delta\varepsilon_\infty^1 \sigma_\nu\right)^2}.$$

(9.16)

Generalizations of the above-mentioned expressions for stresses and their standard deviations can be obtained for the case of anisotropic elastic properties.

### 9.7.6 Summary of FIB-DIC Residual Stress Error Estimation

The steps involved in error propagation in FIB-DIC micro-ring-core analysis are illustrated graphically in the flow diagram in Fig. 9.15, where they are grouped into three main sections: experimentation, DIC analysis of strain relief, and elastic stiffness estimation.

## 9.8 CASE: SEQUENTIAL MILLING FIB-DIC MICRO-RING-CORE RESIDUAL STRESS ANALYSIS IN A SHOT-PEENED NI-SUPERALLOY AEROENGINE TURBINE BLADE

Spatially resolved residual stress analysis offers obvious advantages over single point measurements, through their capacity to reveal stress gradients. This lateral resolution is necessary to improve the understanding of the interactions between microstructure, processing route, and stress state in a range of materials and assemblies (Zhu et al., 2006; Boyce et al., 2001; Sebastiani et al., 2014). To ensure precise knowledge of the stress analysis location and to ensure a consistent gauge volume, a combination of microscopy and measurement is necessary in spatially resolved techniques. For these reasons the microscale ring-core FIB-DIC technique has excellent potential for spatially resolved analysis.

The ring-core FIB-DIC technique introduces annular traction-free surfaces that induce stress (and therefore strain) relief in the surrounding region. Therefore, care must be taken to quantify the distance over which this variation becomes negligible.

To obtain an estimate of this lower limit, calculations on the basis of the classical Lamé thick-walled cylinder solution can be used. Considering the outer

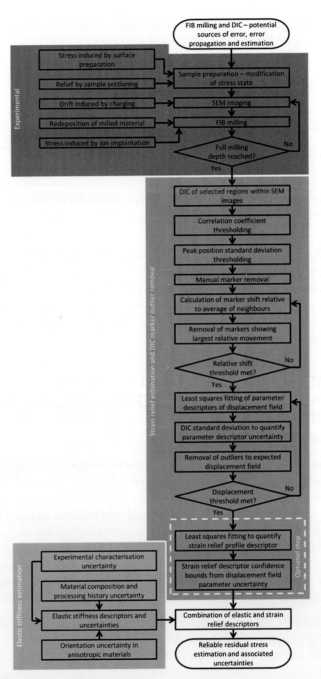

**FIGURE 9.15**
Flow diagram showing potential sources of error and the stages required for error propagation and estimation in Focused Ion Beam (FIB) milling and Digital Image Correlation (DIC) residual stress analysis (Lunt and Korsunsky, 2015). Error sources have been grouped into three categories: experimental, strain relief estimation and DIC marker outlier removal, and elastic stiffness estimation. *SEM*, scanning electron microscopy.

surface of the trench $r_T$ as the inner surface of an infinitely thick cylinder in a state of equibiaxial, uniform in-plane stress $\sigma_R$, the radial stress $(\sigma_{rr})$ and hoop stress $(\sigma_{\theta\theta})$ distributions can be written in a general form as (Sahoo et al., 2013):

$$\sigma_{rr} = C - \frac{D}{r^2}, \quad \sigma_{\theta\theta} = C + \frac{D}{r^2}, \tag{9.17}$$

where the variable $r$ represents the radial coordinate, and $C$ and $D$ are undetermined constants. The traction-free surface at a radius $r_T$ and constant residual stress $\sigma_R$ state at an infinite distance are used as boundary conditions to determine $C$ and $D$ such that

$$\sigma_{rr} = \sigma_R\left[1 - \left(\frac{r_T}{r}\right)^2\right], \quad \sigma_{\theta\theta} = \sigma_R\left[1 + \left(\frac{r_T}{r}\right)^2\right]. \tag{9.18}$$

This simplified stress analysis provides sufficient insight to conclude that the residual stress variation surrounding an annular feature is inversely proportional to the square of the ratio between the radial coordinate and the outer radius of the feature. By comparing the full through depth relief in a thick-walled cylinder and the limited milling depth ($\sim r_T$) of the ring-core FIB-DIC approach, it can be seen that this approximation is an overestimate of the actual stress relief. This means that at a radius equal to five times the island diameter, the induced stress change is guaranteed to fall below 1%.

To provide a quantitative illustration of the capabilities of the *sequential* ring-core FIB-DIC approach, consider an example of material surface response to shot peening. The residual stress profile induced by shot peening has been well characterized by a wide range of previous studies (Sheng et al., 2012; Song et al., 2012a; Torres and Voorwald, 2002; Eberl). The sample selected for the present study was cut from an aeroengine compressor blade made from nickel superalloy IN718. Careful control of the processing route resulted in a microstructure of submicron precipitates of $\gamma'$ phase $Ni_3(Al,Ti)$ within an intermetallic face-centered cubic austenitic phase $\gamma$ matrix. Despite the highly anisotropic nature of grains within Ni-based superalloys, polycrystalline anisotropic approximation was used at the scale of observation ($\sim 5$ μm).

Shot peening was applied to the entire surface of the compressor blade in the direction perpendicular to blade surface. The affected depth was known to be significantly less than 1 mm, placing it beyond the resolution of traditional macroscale techniques. The sequential ring-core FIB-DIC residual stress mapping approach was applied in a line of milled annular features extending from blade surface to $\sim 500$ μm into the bulk.

The sample was cut before analysis to expose a cross-section of the blade. A diamond saw (Buehler Isomet) was used to obtain a blade section, and incremental grinding and polishing was used to minimize residual stresses induced by the preparation process. As a final sample preparation stage, the polished cross-section was etched with Kalling's No. 1 reagent for 60 s. This revealed the $\gamma$ and $\gamma'$ distribution in the microstructure of the underlying material and greatly increased the contrast of the SEM images of the surface.

Ring-core FIB-DIC analysis was performed in the Tescan Lyra 3XM FIB-SEM instrument at the Multi-Beam Laboratory for Engineering Microscopy (MBLEM), Oxford, United Kingdom. Optimization of the SEM parameters was used to generate a spot size of 6.9 nm, and an automated contrast and brightness routine was used to maximize the dynamic range of the captured images. An image size of 2001 × 2001 pixels was selected as a compromise between the resolution and duration of imaging accounting for possible sample drift. Optimization of FIB parameters was performed to generate an effective spot size of 7.5 nm at a beam current of 100 pA selected to reduce ion beam damage.

A pillar diameter of 5 μm was chosen as a balance between milling time (longer for larger diameters) and the precision of stress evaluation (better for larger diameters). A trench width of 1.5 μm was selected to minimize the impact of redeposited material onto the pillar surface and a nominal milling step of 100 nm was used. Following each milling increment, SEM imaging of the core region was performed at an oblique angle of 55°. Tilt correction was used to decrease the vertical scanning increment in the SEM by a factor equal to $1/\cos 55°$, to ensure that the vertical and horizontal imaging resolutions were equal. Milling was performed to a nominal depth of 5.3 μm to ensure that complete stress relief was obtained in the central island. The entire process took approximately 45 min and generated a total of 54 images.

The first milling feature was placed approximately 20 μm from the edge to account for edge rounding following polishing. This process was repeated for 10 points along a line perpendicular to the surface in steps of approximately 50 μm. This spacing was chosen to ensure negligible marker interaction, while providing sufficient resolution to resolve the impact of shot peening. Difficulties in alignment and beam drift meant that some stress analysis points were unsuccessful. For this reason some features were placed closer together and others further apart (Fig. 9.16). The closest markers were placed 25 μm apart corresponding to a maximum potential residual stress deviation of ~2%.

Following SEM image sequence collection, DIC was performed on the core central region, where constant strain variation is expected. Bulk drift was initially removed by performing lower-resolution DIC. Following manual and automated outlier removal, well-tracked markers were used for analysis.

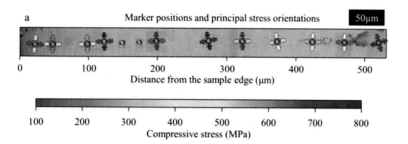

**FIGURE 9.16**

Diagrammatic representation of the magnitude (*arrow color*) and orientation (*arrow rotation*) of the principal stresses over a scanning electron microscopy image of the final milling positions (Lunt and Korsunsky, 2015; Lunt et al., 2015).

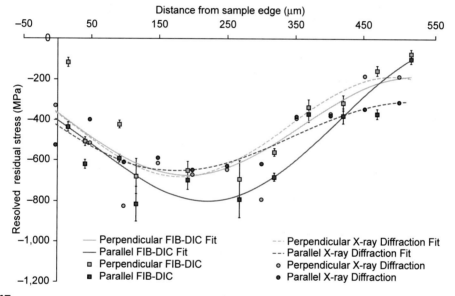

**FIGURE 9.17**

Focused Ion Beam—Digital Image Correlation (FIB-DIC) and X-ray diffraction residual stress estimates in directions parallel and perpendicular to the sample edge, against distance from the sample edge. The error bars indicate the 95% confidence intervals of each measurement (Lunt and Korsunsky, 2015; Lunt et al., 2015).

Plots of displacement ($\Delta x$) against position ($x$) were obtained for each image, and least-squares fitting of a linear profile was used to quantify strain relief ($\Delta\varepsilon = \Delta x/x$) in the $0°$, $45°$, and $90°$ directions. Estimates of the 95% confidence intervals for $\Delta\varepsilon$ were simultaneously obtained from the covariance matrix of this fitting process. The typical output for the feature at $115\ \mu m$ from the surface is shown in Fig. 9.13.

The nominal "master curve" was used to perform least-squares fitting of the strain-depth profiles, obtaining estimates for full strain relief ($\Delta\varepsilon_\infty$). To accommodate minor variations in the milling rate and surface roughness, parameter $\eta$ was used to account for milling rate variation and parameter $\delta$ to account for a minor offset in the milling depth due to surface roughness. The final form of the fitting function used was

$$f(\Delta\varepsilon_\infty, \eta, \delta) = 1.12\Delta\varepsilon_\infty \times \frac{z}{(1+z)}\left[1 + \frac{2}{(1+z^2)}\right], \quad z = \frac{\eta I}{0.42d} + \delta, \quad (9.19)$$

where $d$ is the diameter of the pillar (5 μm). The inverse of the standard deviation of the strain relief estimates were used for data point weighting in the least-squares fitting approach. This enabled accurate estimates to be obtained for the 0°, 45°, and 90° strain relief at infinite depths $\left(\Delta\varepsilon_\infty^{0°}, \Delta\varepsilon_\infty^{45°}, \text{ and } \Delta\varepsilon_\infty^{90°} \text{ respectively}\right)$, as well as the standard deviation of these values.

The principal strain relief orientations, magnitudes, and standard deviations were determined $\left(\Delta\varepsilon_\infty^1 \text{ and } \Delta\varepsilon_\infty^2\right)$ from the 0°, 45°, and 90° strain relief values. The microstructural directionality induced by the shot peening and surface milling meant that the principal directions were closely aligned to the directions parallel and perpendicular to the sample edge, with an average offset of less than 4° (Fig. 9.16).

Principal strain relief values were used to calculate the principal in-plane stress values ($\sigma_1$ and $\sigma_2$) and standard deviations at each marker location, based on the nonequibiaxial stress state assumption. Bulk Young's modulus (205 GPa) and Poisson's ratio (0.294) values for IN718 were used. It is noted that local variations in elastic and plastic anisotropy may affect the stress results obtained. In general, compressive stresses in the range 100–500 MPa were observed near the sample edge. The compressive stress increases to a maximum of $\sim -800$ MPa at $\sim 200$ μm from the sample edge. The magnitude reduces to $\sim 100$ MPa at $\sim 500$ μm from the sample surface.

To validate the results of the sequential ring-core FIB-DIC residual stress analysis, X-ray Powder Diffraction (XRPD) was performed at beamline I15 at Diamond Light Source, Harwell, United Kingdom, using the experimental setup shown in Fig. 9.1. A 70 × 70 μm² collimation assembly was used to define a pencil beam, and photon energy of 76 keV was selected to maximize the incident flux and diffraction signal from the sample.

The sample was placed into a specially manufactured mount and an optical system was used to align the beam with the FIB-DIC marker locations at microscale precision. A raster scan was then used to collect diffraction patterns in increments of 50 μm from the edge of the sample. A Perkin Elmer flat panel 1621-EN detector (2048 × 2048 pixels of size 0.2 × 0.2 mm²) was used to record the resulting diffraction patterns.

A relatively large grain size within the specimen induced graininess in the diffraction patterns. Azimuthal integration over a range of 30° was used to improve the grain sampling statistics of the resulting 1D spectra. A critical examination of the Debye–Scherrer rings revealed that sample graininess had the least impact on the $\gamma$ phase <200> peak. Lattice parameter quantification was performed for the scattering vectors parallel and perpendicular to the sample surface.

To aid in the visualization of the stress distribution and provide comparison with the XRPD results, stresses in directions parallel and perpendicular to the interface were evaluated (Fig. 9.17). Continuous curves provided as a guide to the eye were obtained by fitting the heuristic shot-peened residual stress profile originally proposed by Watanabe et al. (1995):

$$\sigma_R = \alpha[\beta + \gamma z + \{1 + \cos(\theta z + \tau)\}], \qquad (9.20)$$

where $\alpha$, $\beta$, $\gamma$, $\theta$, and $\tau$ are constants, and $z$ is the distance from the edge of the sample. This representation of the residual stress variation is valid up to the limiting depth of the plastic zone; 95% confidence intervals for the resolved FIB-DIC results obtained by the propagation of error values are also shown and correspond to the relative average error of $\sim\pm10\%$. The results of the two techniques follow similar trends, with stresses in the normal direction (labeled "perpendicular") showing a smaller magnitude, particularly toward the surface, in accordance with the traction-free surface requirement.

Joint consideration of the two datasets highlights the challenge of comparing the results obtained using two different techniques. The first point to be noted here concerns the amount of "noise" observed in the two sets of results. It is clear that in the case of XRPD data this is associated with the real uncertainty that arises as a result of the "grainy" nature of 2D diffraction patterns (Debye–Scherrer rings). The difficulty here is associated with the fact that, because of the penetration of high-energy X-rays, and lack of collimation along the incident beam direction, the location of the grains responsible for the strong "spot" reflections remains unknown. In contrast, because FIB-DIC is a surface technique that is associated with high-resolution SEM imaging, the location of the measurement is fully defined to the accuracy of a few microns in 3D. Furthermore, the deviations from the overall trend line displayed by the measured values can be understood in terms of the macro-/microstress classification introduced in Chapter 8. According to this classification, measurements of residual stress conducted at this scale represent the background trend of the macroscopic, millimeter-scale variation outlined by the continuous curves, superimposed on which is the Type II (intergranular) microscopic residual stress that arises because of the inhomogeneity and anisotropy of elastoplastic and thermal properties of individual crystallites. In other words, deviations from the smooth trend seen in the figure are likely not only to contain a limited amount of experimental noise, but also to a significant extent to reflect the underlying microstress variation.

Another aspect of interpretation and cross-correlation between different measurement techniques concerns the mechanical conditions of the gauge volume from which the measurement is obtained. The plane problem of elasticity arises when deformation can be described as 2D, depending only on the in-plane coordinates. Consider the state of plane stress, with axis $x_3$ aligned with the direction of surface normal. In the close proximity of the surface (and absence of steep stress gradients) the out-of-plane stress components $\sigma_{13}$, $\sigma_{23}$, and $\sigma_{33}$ can be neglected. Strains due to the in-plane (residual) stresses can be written as

$$\varepsilon_{12} = \frac{2(1+\nu)}{E}\sigma_{12}, \quad \varepsilon_{11} = \frac{1}{E}[\sigma_{11} - \nu\sigma_{22}], \quad \varepsilon_{22} = \frac{1}{E}[\sigma_{22} - \nu\sigma_{11}] \tag{9.21}$$

The only nonzero out-of-plane strain is found to be $\varepsilon_{33} = -\frac{\nu}{E}(\sigma_{11} + \sigma_{22})$.

*Plane strain* conditions arise if displacements everywhere in a solid body are a function of coordinates in the plane perpendicular to the axis $Ox_3$, while strains $\varepsilon_{13}$, $\varepsilon_{23}$, $\varepsilon_{33}$ vanish. The remaining strains are then given by

$$\varepsilon_{12} = \frac{2(1+\nu)}{E}\sigma_{12}, \quad \varepsilon_{11} = \frac{(1-\nu^2)}{E}\left[\sigma_{11} - \frac{\nu}{1-\nu}\sigma_{22}\right],$$
$$\varepsilon_{22} = \frac{(1-\nu^2)}{E}\left[\sigma_{22} - \frac{\nu}{1-\nu}\sigma_{11}\right]. \tag{9.22}$$

The two sets of equations can be put into equivalent form, provided "plane strain elastic constants" are introduced: $E' = E/(1-\nu^2)$, $\nu' = \nu/(1-\nu)$.

For the purposes of our present analysis we adopt the approximation that cross-section preparation results in the complete relief of out-of-plane residual stresses $\sigma_{13}$, $\sigma_{23}$, $\sigma_{33}$, but (to the first approximation) does not alter the residual elastic strains. This is consistent with the good agreement observed between the XRPD and FIB-DIC measurements.

For a given miller index *hkl*, the conversion between the XRPD lattice parameter ($d_{hkl}$) and the estimate of elastic lattice strain ($\varepsilon_{hkl}$) requires the knowledge of the unstrained lattice parameter $d^0_{hkl}$, as $\varepsilon_{hkl} = \left(d_{hkl} - d^0_{hkl}\right)/d^0_{hkl}$. Thus, to obtain reliable measures of the *absolute* residual stress values, accurate quantification of $d^0_{hkl}$ is essential. Direct comparison between the *absolute* residual stress values obtained by FIB-DIC and the *relative* (i.e., $d^0_{hkl}$ dependent) values obtained by XRPD was used to find the optimal value for the unstrained lattice parameter of face-centered cubic $\gamma$ phase of IN718, $a^0_\gamma = 3.59756$ A, which corresponds well to the existing literature values (Mukherji et al., 2003).

In summary, close agreement is found between the range of stresses (493 MPa and 486 MPa), maximum compressive stress values (680 MPa and 673 MPa), and location (179 μm and 189 μm from the interface) between XRPD and FIB-DIC results, respectively. Grain-level variations in stress were identified as apparent "scatter" around the fitted continuous macrostress profiles. It is noted that the microstress is known to exert a strong influence on crack initiation and propagation, and hence sample failure (King et al., 2008). Owing to its highly spatially resolved nature, micro-ring-core FIB-DIC analysis is identified as a technique capable of capturing the effect of microstress variation.

## 9.9   PARALLEL MILLING FIB-DIC RESIDUAL STRESS ANALYSIS

To increase the spatial resolution of the ring-core FIB-DIC approach, an alternative to the *sequential* approach, based on *parallel* milling of multiple cores, is proposed here. The *parallel* milling approach is based on simultaneously monitoring the strain relief in all cores.

The central parameter of the ring-core FIB-DIC technique is the complete strain relief at the infinite milling depth ($\Delta\varepsilon_\infty$). This is the saturated value of the strain change induced in the surface of the milled island feature and is dependent only upon the residual stress state and the material parameters, and *not* on the milled feature geometry. Although the strain path between the undisturbed and fully relieved state depends on the milling process and the interactions between neighboring features, the total strain change value $\Delta\varepsilon_\infty$ is invariant to these process changes. Continuous imaging of the relief, effective DIC, and the attainment of a sufficient milling depth are all used to obtain the most reliable estimate of $\Delta\varepsilon_\infty$.

A paper by Sebastiani et al. (2014), aimed at quantifying Poisson's ratio at the microscale, demonstrated the impact of alternative milling process routes on the surface relief in an equibiaxial-stressed thin film. Initially, two parallel vertical trenches were FIB milled in the surface (Fig. 9.18A). Following this, two further trenches were milled to leave a 3-μm square "island" of relieved material (Fig. 9.18C). The strain relief variation in the vertical and horizontal directions was then used to estimate Poisson's ratio.

For the purposes of the present discussion we use the results of Sebastiani et al. (2014) to consider the strain relief profiles as a function of depth, as shown in Fig. 9.18D. Although the strain relief profiles in the two directions differ, the strain values converge at large milling depths. This proves that $\Delta\varepsilon_\infty$, is *robust*, i.e., will reach a magnitude that depends only on the undisturbed residual stress. It is known that the *parallel* milling approach will induce smaller

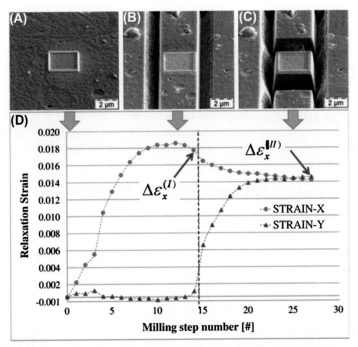

**FIGURE 9.18**

Poisson's ratio determination at the microscale. (A) Digital Image Correlation marker placement. (B) Parallel vertical trench milling. (C) Horizontal milling to leave a 3-μm fully relieved square island (Zhao et al., 2003). (D) The corresponding strain relief curves (Sebastiani et al., 2014).

variations in strain relief than those demonstrated by Sebastiani et al. and that a reliable estimate of $\Delta\varepsilon_\infty$ will therefore be obtained at large milling depths.

This complete stress relief results in an equivalence between the ring-core and square-core approaches, i.e., they will both enable reliable quantification of $\Delta\varepsilon_\infty$. Both techniques can therefore be used interchangeably depending on the specific user shape requirements.

To pursue parallel milling and imaging of multiple cores, a regular line of square features was implemented, the so-called "chocolate block" geometry (Fig. 9.19). This approach ensures that a maximum lateral resolution of one marker width could be reached, with a regular step size between adjacent measurements, and is a much simpler milling regimen than would be required to produce circular markers. Placing the markers close together also reduces the SEM field of view necessary to simultaneously capture all markers, thereby reducing the implementation time. The main limitation on the number of markers comes from the SEM imaging resolution achievable, i.e., the number

**FIGURE 9.19**

Scanning electron microscopy image of the parallel Focused Ion Beam—Digital Image Correlation milling arrangement—the "chocolate block" geometry. Electron deposition of markers has been used to increase the surface contrast of the cores. The average residual stress in the 4-μm cores was determined at an increment of 5 μm (Lunt and Korsunsky, 2015; Lunt et al., 2015).

of pixels that can be captured at the resolution necessary for accurate stress determination.

To guarantee reliable results, the core centers must achieve a state of complete stress relief when milled in a regular arrangement, as proposed in this technique. Synchrotron XRPD mapping (Baimpas et al., 2013) has been used to demonstrate that this is valid for depth to diameter ratios greater than ~0.25 and therefore all milling has been performed to depth–diameter aspect ratios greater than unity.

In terms of the DIC analytical procedure, the *parallel* FIB-DIC approach is very similar to the *sequential* approach at each marker. The main difference is that the strain relief profile is no longer accurately described by the isolated ring-core feature because of the influence of neighboring markers. Nevertheless, accurate estimates of $\Delta\varepsilon_\infty$ can be obtained, provided that the island is milled to a depth sufficient to induce full relief.

## 9.10 CASE: STRESS ANALYSIS IN A CARBON CORE OF A SIC FIBER

Cross-validation between the results of a new experimental technique and a well-established method is a necessary step to assess result reliability. In this regard, a study of the residual strain distribution inside a carbon core silicon carbide (SiC) fiber was selected for comparison (Baimpas et al., 2014b).

In this experiment, XRPD was performed at beamline B16, at Diamond Light Source. High-spatial-resolution (0.4-μm) maps of lattice parameter variation were collected across the carbon core and silicon carbide regions of the uniaxially reinforced titanium alloy (Ti-6Al-4V) composite. To convert this lattice variation into a measure of residual strain, accurate knowledge of the unstrained lattice parameter was required. Insurmountable difficulties arise in producing strain-free powder reference samples of these materials. Therefore, without the high-spatial-resolution (5-µm) analysis performed using the *parallel* FIB-DIC approach, only *relative* information on the strain variation could be obtained.

As previously highlighted, the strain relief obtained during the parallel FIB-DIC approach is a measure of the *absolute* relief in the material surface. Therefore, after performing the necessary XRPD strain value averaging, the unstrained lattice parameters of the SiC and graphite core were obtained by direct comparison between the XRPD and the FIB-DIC strain profiles. Not only did this serve to cross-validate the two experimental techniques, but it also provided the necessary insight to ensure that the nanoscale strain variation determined by XRPD was a measure of the *absolute* strain variation, the critical parameter in understanding the failure modes of these fibers.

At this point, it is important to note that the back-calculation of the residual stress state must be performed with care because of the variations in the amorphous content and associated anisotropy. Nevertheless, the *strain* profiles obtained by XRPD and the FIB-DIC approach can be compared, and show consistent results, as discussed in the following sections.

## 9.10.1   Sample Preparation

The SiC-reinforced titanium alloy composite in this study comprised 35% by volume SCS-6 SiC fibers aligned in a single direction. The fiber was composed of a 30-µm-diameter graphite core that was surrounded by SiC with an outer diameter of 140 µm. Within this graphite core, a distinct untextured central 13.5-µm-diameter region was observed. Two different lattice parameter values were therefore obtained for carbon, one for the inner and one for outer region.

To minimize the residual stresses induced during preparation, sample sectioning was performed using a diamond saw (Buehler Isomet). This was followed by incremental grinding and colloidal silica polishing. A final thickness of $\sim 500$ µm was selected to maximize the diffracted beam intensity at the energies available at beamline B16.

As part of the experimental process outlined in our previous paper (Nelson, 2010), tomographic reconstruction of the SiC was also performed. To facilitate full illumination of the sample by the X-ray beam, further sectioning was performed using a similar diamond saw and polishing process. The final sample was a $1 \times 0.5 \times 0.5$-mm$^3$ cuboid as shown in the insert in Fig. 9.20.

## 9.10.2   X-ray Powder Diffraction Experimental Procedure

To record the highly spatially resolved variations of elastic strain, the KB nanoscale focusing capabilities available at B16 were exploited to produce a $400 \times 500$-nm$^2$ beam. A 150-µm-diameter pinhole was used to "clean up the beam," as shown in Fig. 9.20. The sample was placed on a translation and rotation stage and X-ray imaging was used to align the sample in a

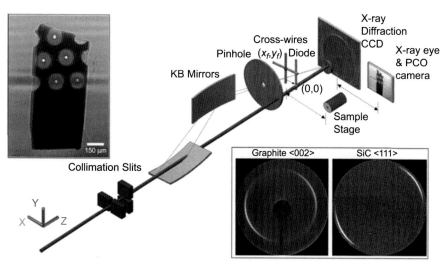

**FIGURE 9.20**

Schematic of the X-ray powder diffraction setup showing the aligned sample and diffraction patterns from the SiC and graphite regions (Baimpas et al., 2014b).

direction parallel to the incident beam. An incremental beam-alignment process was then implemented to determine the location of the beam on the sample surface to nanoscale accuracy; this is outlined in detail elsewhere (Baimpas et al., 2014b).

Piezoelectric translation stages were used to raster the sample incrementally across the beam, and diffraction patterns were recorded at each location. Six line scans were implemented to map a representative region of the SiC and the carbon core.

Azimuthal integration of the resulting diffraction patterns was performed, and the lattice parameter variation was determined for scattering vectors pointing in the radial and hoop directions. The lattice parameter variation in each of the different regions was determined: $a$ in the case of face-centered cubic SiC and $c$, the larger unit cell dimension, for the hexagonal close-packed graphite. As previously noted, two different lattice constants were required in in the graphite: $c_0$ in the case of the outer textured region and $c_i$ in the case of the untextured inner core region. The crystallographic texture associated with the SiC region limited the azimuthal angles over which a representative lattice constant could be quantified and therefore only the variation in the radial lattice constant was determined.

## 9.10.3 The Parallel FIB-DIC Approach

Following XRPD, the sample was placed into the Tescan Lyra 3XM FIB-SEM instrument at MBLEM. SEM parameter optimization was performed to give a

5.7-nm spot size and an image size of $4096 \times 4096$ pixels was selected to maximize the resolution of the captured images. Careful focusing, line and image averaging, and automated contrast and brightness selection were used to maximize the dynamic range and reduce noise in the image.

One of the advantages of the ring-core technique is that during milling the core approaches a state of approximately uniform strain relief. Assuming good image stability, this area averaging minimizes the impact of image distortion on the strain estimate. This means that the ring-core can be reliably implemented at lower magnification, e.g., compared with other FIB-DIC techniques, which rely upon the precise determination of displacement fields, such as hole drilling, or slitting.

Both the SiC and carbon core were found to have similar FIB milling rates and therefore a single parallel milling process was implemented on both regions simultaneously. Six cores with dimensions of $4 \times 4 \ \mu m^2$ were selected as a compromise between maximizing the number of stress evaluation points on one hand, and the precision of these estimates, on the other. A trench width of $1 \ \mu m$ was chosen to minimize the increment between successive points and a depth to diameter ratio of 1.27 was selected to ensure complete relief in the core. The stress analysis positions were located at radial distances between 2.5 and 27.5 $\mu m$ from the fiber axial line, in increments of 5 $\mu m$, as shown in Fig. 9.21.

To overcome the limitations of SEM imaging, a very small FIB milling depth increment of 15 nm was selected. This minimizes the strain change between successive measurements and increases the likelihood of effective DIC tracking. A small milling current of 100 pA was selected to reduce the amount of material redeposition on the islands (and the associated image blurring in the DIC analysis) and to minimize the residual stress induced by gallium ion implantation.

The optimized arrangement captured 340 images over a period of approximately 5 h. Although this time period may seem long, the full 2D in-plane stress state is characterized at six different locations during this interval. Eighteen independent implementations of a 1D stress characterization technique would be necessary to obtain comparable data, resulting in an equivalent time budget of 17 min per data point for the *parallel* FIB-DIC method.

DIC of the resulting images was performed using a modified version of the DIC script developed by Eberl et al. It was found that the residual colloidal silica provided increased surface contrast thereby improving marker tracking effectiveness. Bulk drift was initially accounted for by performing lower-resolution DIC, and each core surface was tracked individually during six

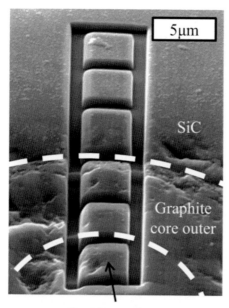

Graphite core inner

**FIGURE 9.21**

Scanning electron microscopy image of the parallel Focused Ion Beam—Digital Image Correlation
technique in which the interfaces between the SiC region, the graphite core outer, and the graphite core
inner are highlighted. The core sizes are uniform in the direction of the global surface normal. However,
slight variations in topology create the illusion of size variation. The Z-contrast induced by colloidal silica is
also shown (Baimpas et al., 2014b).

independent implementations of the script. Automated and manual outlier
removal was used to retain only the well-tracked markers.

In this study, strain relief profiles were obtained in the radial and hoop directions for each of the six cores. Although theoretically possible, the full in-plane
strain tensor was not quantified. This is because unknown mechanical property
variations prevent the conversion of the strain relief results into stress. Therefore, it was decided that full tensor analysis would offer limited value for the
present study.

Relatively high levels of noise (and the associated 95% confidence intervals)
were observed in the profile of strain relief against the image number (i.e., milling depth). A weighted average of multiple (20) markers was therefore calculated based on the inverse of the standard deviation of each relief value.
Following normalization against the full-depth strain relief values, the results
of this averaging are shown in Fig. 9.22. The image number of each point
was chosen as the central image number over which the strain relief values
had been averaged.

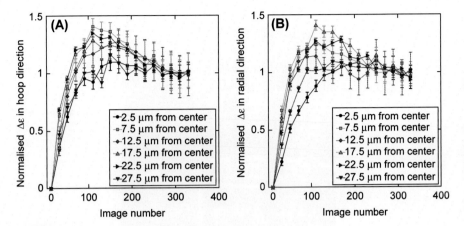

**FIGURE 9.22**

Normalized strain relief 20-point weighted average against image number for the six stress analysis locations, in the hoop (A) and radial (B) directions (Lunt and Korsunsky, 2015; Lunt et al., 2015).

The first conclusion drawn from Fig. 9.22 is that milling has been performed to a depth sufficient to induce full relief in the cores. This is demonstrated by the plateau observed in all strain relief profiles.

The influence of neighboring markers is also demonstrated in Fig. 9.22, through the slight differences observed in the strain relief profiles. Feature symmetry suggests that the profiles at 2.5 and 27.5 μm, 7.5 and 22.5 μm, and 12.5 and 17.5 μm should be similar. Careful examination of the profiles confirms the presence of a degree of similarity between these profiles.

Despite minor variations in the strain relief profiles, the general variation is very similar to the functional form of the isotropic, single ring-core strain relief master function. Based on this insight, careful least-squares fitting to the data was performed using this function. It is believed that minor variations in the near-surface relief profile should have a limited impact on the estimate of $\Delta\varepsilon_\infty$ obtained from this process. This is increasingly true for profiles for which a large milling depth has been reached. To provide additional support for this assumption, FE modeling of the *parallel* FIB-DIC feature geometry could be used. However, the authors are confident of the reliability of the approach, both from the theoretical point of view and experience. This conclusion is supported by the agreement observed between the XRPD and FIB-DIC results.

## 9.10.4 Experimental Results

To interpret the SiC and graphite XRPD lattice parameter variation in terms of strain, accurate quantification of the unstrained lattice parameters was required. A least-squares optimization approach was therefore implemented starting from the literature values of the unstrained lattice parameters

$[a^0 = 4.3596$ Å (Henson and Reynolds, 1965) for the face-centered cubic SiC region and $c^0 = 6.720$ Å for the larger unit cell dimension of the graphite region]. For this analysis, averaging of the XRPD strain values over the relevant gauge volume was necessary to provide comparative values. For example, the FIB-DIC estimate at the 2.5-μm position represents a strain average between the radii of 0.5 and 4.5 μm. The optimized values were found to be $a^0 = 4.3982$ Å, $c_o^0 = 6.8353$ Å, and $c_i^0 = 6.9980$ Å, where the superscript 0 refers to the unstrained lattice parameters of the SiC ($a$), graphite outer region ($c_0$), and the graphite inner region ($c_i$).

The variations in the *absolute* residual strain obtained from the XRPD and FIB-DIC approach are plotted together in Fig. 9.23. The SiC region can be seen to be in a state of compressive strain that decreases in magnitude with distance from the core. The central carbon core region is in a state of approximate hydrostatic compressive strain of ~0.4%. The outer carbon core, on the other hand, is in a state of dilatational strain; the radial component is tensile and the hoop strain is compressive.

## 9.10.5   Discussion

To calculate the confidence intervals of the FIB-DIC strain relief values, careful error propagation was performed through the multiple stages of least-squares fitting. Taking into account the expected differences between the strain relief profiles (Fig. 9.22), functional fitting ensured that relatively large 95% confidence intervals were obtained, with a typical uncertainty measure of ±20%.

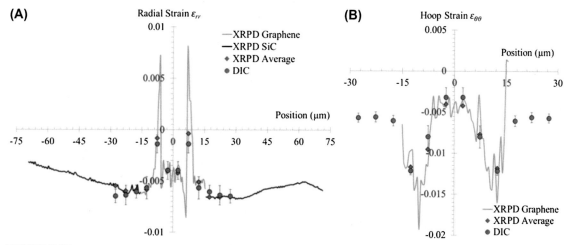

**FIGURE 9.23**

Radial (A) and hoop (B) absolute strain distributions within the graphite and SiC regions of the core (Baimpas et al., 2014b). The results of both the X-ray powder diffraction (XRPD) and the parallel Focused Ion Beam—Digital Image Correlation (DIC) are shown, as well as the comparable average XRPD results. The 95% confidence intervals of the XRPD values are indicated by the error bars.

Examination of the average XRPD data reveals that only 3 of the 18 data points fall outside the FIB-DIC 95% confidence bounds, and that the average percentile error is 13%. Considering that a number of assumptions need to be made to compare different techniques, the two data sets show strong similarities in the distributions obtained, confirming that the *parallel* spatially resolved FIB-DIC approach is a reliable method for in-plane absolute residual distributions of strain and stress in well-characterized materials.

The benefits of the *parallel* FIB-DIC approach can be demonstrated by a critical comparison with high-spatial-resolution analysis of residual stress using slotting-based methods that employ a single slot to determine residual stress variation along the slot in the direction perpendicular to the slot. Although this is a marginally simpler experimental form, the requirement of repeated FE simulations and full displacement field characterization means that the processing time and its complexity exceed those in the *parallel* FIB-DIC approach. The full-field characterization approach also means that strain averaging cannot be performed, and therefore the impact of noise is larger. Furthermore, the impact of edge effects at the end of the slot greatly reduces the precision of the slot-based technique in these regions. Most importantly, the ability to determine the full in-plane stress tensor using the *parallel* FIB-DIC approach, rather than the 1D stress state, enables a much greater insight into the likely failure modes or stress interaction in the region of interest.

Despite the advantages of the *parallel* FIB-DIC technique its use is likely to be limited to specialized applications where micron-resolved residual stress analysis is crucial in improving current understanding. This restriction is primarily associated with the long milling times ($\sim 5$ h), although dramatically shorter milling times would be possible if less precise measurements of residual stress are required (increased milling rates typically reduce DIC accuracy). Other restrictions on the technique include the minimum resolution and maximum region over which stress can be assessed, very-high-resolution analysis ($<1$ μm) would likely begin to be influenced by the effects of ion implantation and reduced DIC areas, whereas limitations on the maximum high-resolution SEM image sizes and increased milling times place an upper boundary on the assessment region. Finally, as highlighted in the carbon fiber example study, difficulties in obtaining precise and reliable stiffness tensor matrices at these resolutions may provide challenges in the conversion of the strain relief values to residual stress estimates.

# CHAPTER 10

# The Inverse Eigenstrain Method of Residual Stress Reconstruction

*Marina Tsvetaeva: Monologue 1913*
*(AMK translation in Appendix C)*

*Pietro Mascagni, Intermezzo from Cavalleria Rusticana, 1890*
*Gustavo Dudamel and Goteborgs Synfoniker*

*Pablo Ruiz Picasso, Self-Portrait (oil on canvas, 24 × 33 cm)*
*1896 Picasso Museum, Barcelona*

## 10.1 FUNDAMENTALS OF INVERSE EIGENSTRAIN ANALYSIS

The class of problems that we presently wish to address stands in an inverse relationship to those discussed in the early chapters. We start from the position of possessing limited knowledge of the residual elastic strain (RES) distribution, e.g., from diffraction measurement. Alternatively, increments of the elastic strain values could have been monitored, e.g., using strain gauges, in the course of material removal. The underlying eigenstrain distribution is to be determined.

In practice the RES, or its increments, can be measured only at a finite number of points. We are therefore seeking to *reconstruct* an unknown functional distribution, i.e., an object with infinite number of degrees of freedom, using a finite data set. Several difficulties may arise in this situation, e.g., whether the problem described in the previous section *can* be inverted; whether the inverse problem is regular, i.e., varies in a smooth manner depending on the data; and whether the obtained solution is unique. In the present study we do not attempt to answer these questions. Instead, we offer an efficient inversion procedure, leaving the evaluation of its uniqueness and regularity for future consideration.

Consider a set of experimental data consisting of the values of RES $y_j$ collected at positions $x_j$, $j = 1, \ldots, m$. In the present study we assume that the data were collected from a one-dimensional scan in coordinate $x$. It is worth noting,

## CONTENTS

A Teaching Essay on Residual Stresses and Eigenstrains. http://dx.doi.org/10.1016/B978-0-12-810990-8.00010-0

however, that the approach presented later is not in any way limited to one-dimensional problems, and can be readily generalized to two- and three-dimensional cases.

Denote by $e(x)$, as in the previous section, the *predicted*, or *modeled*, RES distribution. Evaluating $e(x)$ at each of the measurement points gives the *predicted values* $e_j = e(x_j)$. To measure the goodness of the prediction we form a functional $J$ given by the sum of squares of differences between the actual measurements and the predicted values, with weights:

$$J = \sum_{j=1}^{m} w_j \left( y_j - e_j \right)^2. \tag{10.1}$$

The choice of weights $w_j$ is left to the modeler; for example, they could be chosen based on the accuracy of the measurements being interpreted.

Minimization of functional $J$ provides a rational variational basis for selecting the most suitable model to match the measurements, in terms of the overall goodness of fit.

Let us now assume that the unknown eigenstrain distribution, yet to be determined, is given by a truncated series of *basis* distributions,

$$\varepsilon^*(x) = \sum_{i=1}^{N} c_i \xi_i(x). \tag{10.2}$$

Here $N$ is the total number of *basis* distributions used in the prediction.

An analytical procedure has been established previously for the solution of the *direct problem*, i.e., the determination of the RES distribution that arises in response to an arbitrary eigenstrain distribution $\varepsilon^*(x)$. This procedure can now be applied to each of the $N$ basis distributions $\xi_i(x)$ in turn. As a result, a family of RES solutions $E_i(x)$ is obtained.

Owing to the linearity of the direct problem, the predicted values $e_j$ of the RES due to the eigenstrain distribution $\varepsilon^*(x)$ of Eq. (10.2) can themselves be written in the form of a superposition of responses to the *basis* eigenstrain distributions,

$$e_j = \sum_{i=1}^{N} c_i E_i(x_j) = \sum_{i-1}^{N} c_i e_{ij}, \tag{10.3}$$

with the same coefficients $c_i$ as in Eq. (10.2).

The inverse problem of determining the unknown eigenstrain distribution $\varepsilon^*(x)$ has now been reduced to the problem of determination of $N$ unknown coefficients $c_i$ that deliver a minimum to the functional $J$ in Eq. (10.1), which may now be rewritten as

$$J = \sum_{j=1}^{m} w_j \left( \sum_{i=1}^{N} c_i e_{ij} - y_j \right)^2 .$$

(10.4)

This expression is quadratic and positive definite in the unknown coefficients $c_i$. It follows that the functional has a unique minimum that is found by satisfying the condition

$$\nabla_c J = 0, \quad \text{or } \partial J / \partial c_i = 0, \ i = 1, \dots, N.$$

(10.5)

Owing to the quadratic nature of the functional in Eq. (10.4), the system of equations in Eq. (10.5) is linear. Therefore, the solution for the unknown coefficients $c_i$ can be readily found *without iteration* by inverting the linear system arising in Eq. (10.5). This system is written out explicitly later.

The partial derivative of $J$ with respect to the coefficient $c_i$ can be written explicitly:

$$\partial J / \partial c_i = 2 \sum_{j=1}^{m} w_j e_{ij} \left( \sum_{k=1}^{N} c_k e_{kj} - y_j \right) = 2 \left( \sum_{k=1}^{N} c_k \sum_{j=1}^{m} w_j e_{ij} e_{kj} - \sum_{j=1}^{m} w_j e_{ij} y_j \right) = 0.$$

(10.6)

For purposes of illustration, let us now assume that the weights are equal to unity, so that Eq. (10.6) simplifies to

$$\partial J / \partial c_i = 2 \left( \sum_{k=1}^{N} c_k \sum_{j=1}^{m} e_{ij} e_{kj} - \sum_{j=1}^{m} e_{ij} y_j \right) = 0.$$

(10.7)

We introduce the following matrix and vector notation:

$$\mathbf{E} = \{e_{ij}\}, \quad \mathbf{y} = \{y_j\}, \quad \mathbf{c} = \{c_i\}.$$

(10.8)

Noting that notation $e_{kj}$ corresponds to the transpose of matrix $\mathbf{E}$, the entities appearing in Eq. (10.7) can be written in matrix form as

$$\mathbf{A} = \sum_{j=1}^{m} e_{ij} e_{kj} = \mathbf{E}\mathbf{E}^T, \quad \mathbf{b} = \sum_{j=1}^{m} e_{ij} y_j = \mathbf{E}\mathbf{y}.$$

(10.9)

Hence Eq. (10.7) assumes the form

$$\nabla_c J = 2(\mathbf{A}\mathbf{c} - \mathbf{b}) = 0.$$

(10.10)

The solution of the inverse problem has thus been reduced to the solution of the linear system

$$\mathbf{A}\mathbf{c} = \mathbf{b}$$

(10.11)

for the unknown vector of coefficients $\mathbf{c} = \{c_i\}$.

Whenever the solution of an inverse problem is sought, questions arise concerning the existence and uniqueness of the solution, and also concerning the well-posedness of the problem, i.e., the continuity of the solution dependence on the problem parameters, the choice of the basis functions, the number of terms $N$ in the truncated series, etc.

Within the present regularized formulation of the problem, for an *arbitrary* choice of the family of basis functions and *arbitrary* number of basis functions $N$, a unique solution is guaranteed to exist. This is a consequence of the positive definiteness of the quadratic functional $J$. Furthermore, it is clear that increasing the number of terms $N$ is guaranteed to deliver a sequence of monotonically nonincreasing values of $J$, i.e., the goodness of approximation will not be diminished.

An interesting question concerns the convergence of the solution, e.g., in terms of eigenstrain distribution, to the "true" solution, in the limit. Similarly, the continuity in the behavior of the solution with the choice of basis functions deserves to be discussed. Although it must be emphasized that these questions are clearly fundamental and ought to be addressed, the focus is currently placed on the development of a practical tool for residual strain analysis. In so far as this is the aim of the present study, the proposed framework offers an efficient "one shot" approach to the solution of inverse problem. Furthermore, the choice of moderate values $N$, compared with the number of measurements, $m$, offers a rational procedure for smoothing the data, as discussed in the next section.

## 10.2   INVERSE EIGENSTRAIN ANALYSIS OF AN INELASTICALLY BENT BEAM

To illustrate the approach to solving the inverse problem of eigenstrain we return to the case of the residually bent beam. An illustration of the application of the inverse eigenstrain procedure to the interpretation of experimental diffraction data is shown in Fig. 10.1.

The data for RES (plotted using the units of microstrain, $10^{-6}$) obtained by diffraction are shown by the markers. The light dashed line denoted as "Model LIN" represents the reconstruction obtained using a simple eigenstrain approximation consisting of two linear segments at the beam extremes of the type illustrated in Fig. 4.4, the slopes of the two segments being the variables sought in the inverse eigenstrain analysis. The heavy dashed curve denoted as "Model NLIN" corresponds to the result of the eigenstrain reconstruction using a power law series representation that achieves a closer agreement with the measurements.

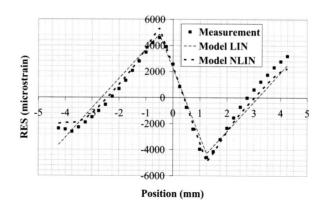

**FIGURE 10.1**

Illustration of application of the inverse eigenstrain procedure. The measured profile of residual elastic strains (RES) in the bent Ti-6Al-4V bar (*markers*) is compared with the predictions of the linear eigenstrain model (*thin dashed line*) and of the higher-order eigenstrain bending model (*thicker dashed line*). (Korsunsky, 2006a).

## 10.3 INVERSE EIGENSTRAIN ANALYSIS OF WELDS

Residual stresses in welded structures are frequently of particular concern in design, because these joints often act as the sources of fracture initiation leading to failure. Reliable prediction of residual stresses associated with welds is a notoriously challenging problem that has motivated researchers to seek reliable experimental means of stress state evaluation. The difficulty that arises in this context is that stress evaluation can be accomplished only at a limited number of points, whereas obtaining the complete picture is desired.

Inverse eigenstrain analysis offers a powerful means of addressing this challenge. Let us suppose that the RES has been evaluated at a finite number of points, for example, by diffraction. The inverse problem formulation introduced previously allows the underlying eigenstrain field to be sought that acts as the source of the residual stress that provides optimal match to the observations.

Fig. 10.2 illustrates the quality of matching that can be achieved using the inverse eigenstrain approach that was applied using the Chebyshev polynomial representation of eigenstrain variation across the weld.

## 10.4 INVERSE EIGENSTRAIN ANALYSIS OF LASER SHOCK-PEENED SAMPLES

Laser shock peening is a technique that allows the introduction of compressive residual stresses into components to the depths in excess of 1 mm. The mechanism of residual stress generation is similar to conventional shot peening, but

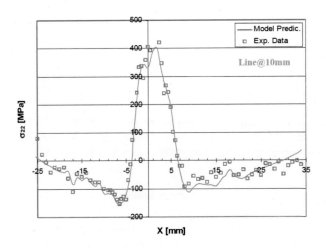

**FIGURE 10.2**

Inverse eigenstrain reconstruction (*continuous curve*) of residual stress variation along a line crossing the weld at $x = 10$, obtained by matching the measurements of residual elastic strains indicated by the *markers* (Korsunsky et al., 2007).

plastic compression of the surface layers is accomplished by the passage of a shock wave induced by laser ablation of a sacrificial layer of materials (or water) at the sample surface.

It is apparent in this case that the RES distribution can be thought of as consisting of two principal components: the near-surface eigenstrain profile shown in Fig. 10.3B and the equilibrating elastic profile due to bending. These combine to give rise to the overall RES profile shown in Fig. 10.3A. The quality of fit is limited by the simplistic form of eigenstrain assumed for the purpose of simple analytical treatment in the form of a shifted portion of a cosine function. More careful choice of the eigenstrain representation can deliver an even better match, although it is important to note that some of the fine-scale features observed in the profile are likely to be associated with material processing operations before laser shock treatment (e.g., plate rolling).

## 10.5 STRAIN TOMOGRAPHY AND RELATED PROBLEMS

Strain tomography is an example of a broader concept of "rich tomography," an approach aimed at the reconstruction of three-dimensional variation of a complex nonscalar quantity (vector, tensor, distribution) based on two-dimensional projection measurements. This problem falls into the category of inverse problems for which considerations of existence and uniqueness of solution play a particularly important role, as do the sensitivity and stability of the solution. Conventional computed tomography performs reconstruction

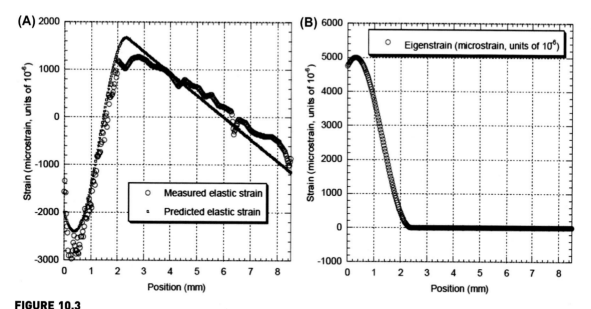

**FIGURE 10.3**

(A) Inverse eigenstrain reconstruction of residual strain in a laser shock-peened sample, and (B) the underlying eigenstrain distribution (Korsunsky, 2006b).

of a scalar, most commonly the absorption coefficient, based on a redundant dataset obtained from multiple registration of transmitted beam intensity at various angles of projection. Under certain not particularly restrictive conditions this problem is known to be well posed, allowing the reconstruction by filtered back-projection (inverse Radon transform).

However, when the quantity of interest is multicomponent, some elements may become "invisible," preventing the application of filtered back-projection. In the context of X-ray science, multiple attempts have been made to replace imaging mode with diffraction, seeking to gain a route to accessing important structural characteristics such as crystal orientation and lattice strain. It is evident, however, that the diffraction condition may not be satisfied for all angles of rotation, and some components of strain may not be measureable. If the component of interest is invariant with respect to sample rotation (e.g., is parallel to the axis), then a scalar problem formulation can be used, and the solution obtained (Korsunsky et al., 2007). The crucial steps in the adaptation of diffraction strain tomography to reconstruction by conventional filtered back projection involve (1) proving that the strain deduced from diffraction from an extended gauge volume corresponds to the average value within this volume, and (2) the evaluation of the gauge volume length along the incident beam for each sample position and orientation angle, so that the average value can be converted to the integral required as input for inverse Radon transform.

A possible route to overcoming invisibility is to use white beam instead of monochromatic, ensuring that a Laue scattering pattern is always obtained. For the case of crystal orientation that can be described by Euler angles or Rodrigues vector, it has been demonstrated that three-dimensional reconstruction of grain orientation within a bulk sample (and grain misorientation within individual grains) can be achieved this way (Baimpas et al., 2014a).

A particularly interesting challenge in strain tomography arises in connection with the use of neutron Bragg edge imaging (Kockelmann, 2007). The nature of the experimental setup in this case is such that the features (edges) present in the energy spectrum of transmitted neutrons contain information about the average lattice parameter (and hence strain) in the direction defined by the incident beam. The corresponding strain value therefore corresponds to the average direct strain in this direction within the sample. In combination with the high penetration ability of neutron beams this offers a powerful potential route to bulk residual stress reconstruction. However, it is easy to see that the information about strains in the plane perpendicular to the axis of rotation is incomplete. For example, the strain component in the direction normal to the sample boundary makes no direct contribution to the measured average value, and therefore cannot be backed out using conventional tomographic reconstruction. An alternative approach can be developed that takes into account the mechanical constraints to regularize the inverse problem formulation. For example, the surface traction at the sample boundary must vanish, which imposes an additional condition on the distribution of RES. An even more robust formulation can be obtained if *eigenstrain* is chosen as the

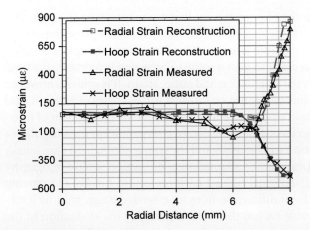

**FIGURE 10.4**

Illustration of residual elastic strains in a quenched cylinder measured experimentally by neutron diffraction, shown together with their reconstruction by neutron strain tomography (Abbey et al., 2012).

principal unknown field. First, some a priori information about eigenstrain may be available from the knowledge of the sample processing history. Furthermore, unlike RES, eigenstrain does not need to correspond to a self-equilibrated stress field, conferring greater flexibility on the inverse problem formulation.

An example of residual strain reconstruction in a quenched cylinder is shown in Fig. 10.4. The limited degree of agreement with the predictions of the simulation described in Section 5.3 reveals the likelihood of reverse plastic flow during cooling.

# CHAPTER 11

# Eigenstrain Methods in Structural Integrity Analysis

*Virginia Woolf: Uncle Vanya 1921*

*Anna Netrebko/Rolando Villazon, O soave fanciulla*
*La Bohéme, Giacomo Puccini, 1896*

*Valentin Serov, Girl with peaches (oil on canvas, 85 × 91 cm) 1887*
*Tretyakov Gallery, Moscow*

A persistent challenge arises in the context of structural integrity analysis, particularly for the purpose of estimating safe fatigue life of components and assemblies that were produced by complex thermomechanical processing. This challenge arises from the fact that the current state of the object (specimen, component) needs to be known in sufficient detail to enable subsequent numerical modeling of its evolution in the course of its service operation. Consider as an example a blade of an aeroengine turbine, which in service experiences time-dependent thermal and mechanical loading. Although the existing methods for the prediction of crack initiation and propagation in such components are sufficiently well advanced to provide confident predictions of safe service lives, it is also well known that fatigue processes are strongly dependent on the object's spatially distributed residual stress and microstructural state—*current state*—which needs to be taken into account. Although the methods for reliable visualization and description of the microstructural state have been advanced over centuries, suitably spatially resolved methods of residual stress analysis have emerged only towards the end of the 20th century. Moreover, how can one incorporate residual stress information in a numerical model, given the complex, multicomponent, three-dimensional nature of this parameter, which needs to satisfy the requirements of equilibrium and boundary conditions, and can only ever be measured at a finite number of locations with limited resolution?

One approach to addressing this challenge is via process modeling: the entire processing history of the object needs to be simulated numerically, taking into account each processing stage, to arrive at a numerical model of the current state. If the object experienced rolling, forming, diffusion bonding, annealing

## CONTENTS

**167**

A Teaching Essay on Residual Stresses and Eigenstrains. http://dx.doi.org/10.1016/B978-0-12-810990-8.00011-2

heat treatment, surface modification by cold rolling, and shot peening, etc., then each of these stages must be adequately captured in the process model. Moreover, to arrive at reliable predictions, time- and temperature-dependent material parameters must be known, e.g., thermal conductivity, hardening behavior, solute concentration, grain size, and texture, at each stage of processing! Researchers have mounted a concerted attack on this major problem, undaunted by the magnitude of the task, ready to model each instance of impact of a single shot on the sample surface to predict the residual stress state that arises as a consequence, but even given such heroic efforts, it is difficult to expect quantitatively reliable results to emerge, given the multitude of tabular parametric data that need to be made available to them.

An alternative approach to this challenging problem can be proposed, which we term *current state modeling* here. The idea of this approach lies in combining the characterization of the object in its present form (in terms of microstructure, texture, residual stress), with minimal reference being made to its processing history, with formulating a mechanistic model of its status that can serve as the basis for simulating its behavior in subsequent service. The incorporation of microstructure data into numerical models in the form of spatially distributed internal variables is well established in the context of finite elements and other numerical methods. Mechanical properties such as stiffness, strength, hardening behavior, and fracture toughness can be obtained from spatially resolved *mechanical microscopy* experiments, e.g., micropillar compression and splitting by nanoindentation.

The final bit of the puzzle is residual stress that needs to be evaluated at multiple scales (Type I-II-III). Even given the very best tools and unlimited access to them, the data collected will inevitably be limited, yet defining a global current state model is required. In this situation, the *inverse eigenstrain* problem formulation comes to the rescue: given even a limited dataset pertaining to the residual stress state within the object, it allows the reconstruction of the global, multicomponent, multiscale description of the internal stress state that is consistent with the mechanical requirements, and represents a most probable match to available observations that can incorporate any a priori information about the nature of prior processing, e.g., by restricting the domain to which plastic deformation (and hence eigenstrain) must be localized.

Current state modeling is a relatively new idea that requires further development. We present here a few examples that may help to illustrate this approach.

## 11.1 EIGENSTRAIN ANALYSIS IN TRIBOLOGY

Fretting fatigue is the phenomenon of crack initiation and propagation from the region of stress concentration in reciprocating frictional contacts. It has been identified as the cause of serious damage in dovetail connections of aero-space engines. Shot peening is a well-known method used to improve fatigue resistance of steels, aluminum alloys, nickel alloys, and titanium alloys, both for structural elements subjected to cyclic tensile loading and for contacting surfaces.

Shot peen treatment is frequently used as a means of protecting sample surfaces from crack initiation. It involves the use of peen medium in the form of hard-ened steel balls or glass beads (shot) that is entrained in a fast flow of air and impacts on the surface with momentum that is sufficient to create a shallow re-gion of plastic deformation. Under normal impact conditions, the induced plastic strain is compressive in the direction of the surface normal (say, $z$ axis) and equibiaxially tensile $\left(\varepsilon_x^* = \varepsilon_y^*\right)$ in the in-plane $x$ and $y$ directions. Because plastic deformation is volume conserving, the total volumetric strain equals zero:

$$\Delta = \varepsilon_x^* + \varepsilon_y^* + \varepsilon_z^* = 0. \tag{11.1}$$

Because $\varepsilon_x^* = \varepsilon_y^*$, Eq. (11.1) implies that the three principal components of eigenstrain are related as follows:

$$\varepsilon_x^* = \varepsilon_y^* = -2\varepsilon_z^* = f(z). \tag{11.2}$$

Here $f(z)$ is the function that describes the depth variation of permanent in-plane tensile eigenstrain. Because shot peening induces permanent deforma-tion only within a shallow layer that does not exceed the shot radius, function $f(z)$ is of finite support, i.e., it assumes nonzero values only within the region $0 < z < z_{max}$, where $z_{max}$ is the maximum depth at which plastic deformation takes place.

Fig. 11.1 shows a schematic diagram of an experimental setup for fretting fa-tigue testing of shot-peened samples. In the experiment, two opposing flat-and-rounded pads simulating the conditions arising in dovetail connections in aeroengine fan assemblies are pressed against a bar specimen that is sub-jected to cyclic tension. This causes relative displacement in the contacting pairs in the stick-slip regime and creates fretting conditions at the edges of contact.

In the study by K. Kim (2005), X-ray diffraction experiments were used to evaluate the residual elastic strains (RES) in the shot-peened bar sample before and after fretting. The RES before fretting were used for inverse eigenstrain determination that allowed the formulation of a finite element model. The model was then used to simulate the residual stress "shakedown" due to the sliding contact at the edge of bedding, and the prediction was compared with the residual stress based on RES measurements performed on the sample after fretting (Fig. 11.2).

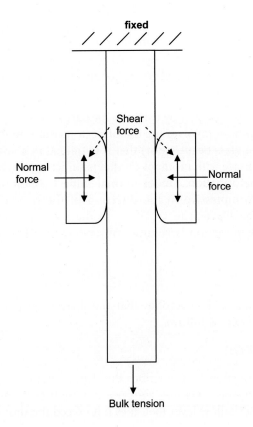

**FIGURE 11.1**

Illustration of the fretting fatigue experiment using flat-and-rounded pads against a sample subjected to cyclic bulk tension (Kim, 2005).

It is apparent from Fig. 11.2 that the combination of inverse eigenstrain reconstruction of residual stress due to shot peening before fretting with the process modeling of residual stress modification due to additional plastic deformation arising during fretting leads to satisfactory agreement with the observations. This illustrates that the eigenstrain approach can serve as a powerful tool for model validation.

## 11.2 EIGENSTRAIN ANALYSIS IN FRACTURE MECHANICS

Plastic deformation and residual stresses arising as a result of the propagation of a fatigue crack play a crucial role in determining the rate of crack growth. For example, the combination of fatigue crack closure and compressive stress ahead of the crack tip that arises as a consequence of overload causes considerable

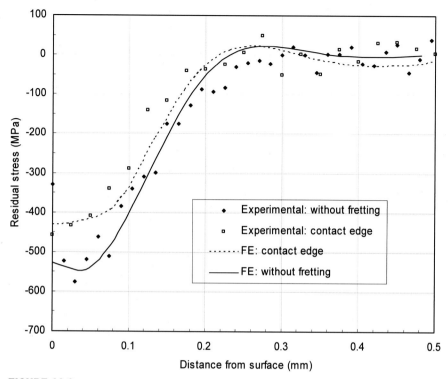

**FIGURE 11.2**

Comparison of residual stress in the sample before fretting (*solid markers* and *continuous curve*) that served as the basis for finite element simulation, with the prediction (*dashed curve*) and measurement results (*open markers*) after fretting (Kim, 2005).

crack retardation. Understanding this effect, along with the crack acceleration following the application of compressive underload, represents an important step toward robust and reliable modeling of complex "spectrum" loading that is often encountered in service, in contrast with the simplified constant amplitude fatigue analysis that is often used as the basis for life prediction.

Eigenstrain analysis can be used to good effect to describe these phenomena, provided reliable information is available to set up inverse eigenstrain analysis with the purpose of eigenstrain determination. Early attempts to follow this approach were hampered by the lack of a residual stress evaluation procedure that offered the required combination of precision and resolution. Multiple attempts have been made to perform the measurements using X-ray diffraction. However, despite the improving spatial resolution of this technique, additional difficulties often arise in analyzing bulk samples because of crack front curvature. The advent of micron resolution Focused Ion Beam–Digital Image

Correlation (FIB-DIC) microring core improved the outlook for such studies. Salvati et al. (2017) reported that the effects of crack closure and residual stress on crack retardation following an overload could be separated, indicating that the quality of measured data was sufficient to extract detailed information about the conditions of crack growth. Further work in this field can proceed by employing parallel FIB-DIC analysis to back out microscale eigenstrain distributions in the vicinity of crack tips. Combined with the characterization of material plasticity and hardening at the microscale by nanoindentation this opens the way toward performing eigenstrain-based analysis of crack propagation using an approach similar to that described in the previous section.

# CHAPTER 12

# Conclusions and Outlook

*J. Galsworthy, Interlude: Indian Summer of a Forsyte, 1918*

*M. Caballé, Casta diva*
*Vincenzo Bellini, Norma 1831*

*Ivan Kramskoy, Portrait of Leo Tolstoy (oil on canvas 80 × 98 cm) 1873*
*Tretyakov Gallery, Moscow*

What conclusions can be drawn from the patchy picture painted in this essay?

The subject of experimental and modeling residual stress analysis has made great strides in the early decades, moving from the level of conceptual schematic diagrams for scale-based residual stress classification to the present state, in which a selection of versatile, high-resolution evaluation techniques is available, along with advanced tools for numerical simulation. Importantly, this has been accompanied by the development and refinement of fundamental approaches based on the concept of *eigenstrain* that has provided a unifying and scale-independent basis for residual stress analysis.

The subject remains largely open-ended: new approaches are being proposed; new techniques uncover the possibilities of analysis at ever finer resolution, all the way down to the atomic scale; and, importantly, new problems are identified in which residual stresses play an intricate yet crucial role in determining not only the structural integrity, but also the functional performance of components and assemblies. This concerns electronic devices in which through-silicon-via (TSV) integration requires the combination of materials with distinctly different coefficients of thermal expansion, leading to residual stress generation that has serious implications for design. The progress in micro- and nanoelectromechanical systems brings with it an interesting set of challenges related to the effects of residual stresses on vibration properties and reliability of operation under the conditions of very high cycle fatigue.

A Teaching Essay on Residual Stresses and Eigenstrains. http://dx.doi.org/10.1016/B978-0-12-810990-8.00012-4

In my opinion, it is the combination of these micro- and nanoscale challenges with the advancement of characterization tools toward high-resolution, high-accuracy measurements that will constitute the main thrust in residual stress analysis in the coming decade.

It will be an exhilarating journey, and I hope that by this point the reader may find that it might be a good idea to bring along this short book as a companion.

# Appendix A: The Eshelby Solution

The Eshelby closed form solution provides the expressions for the elastic fields inside and outside an ellipsoidal inclusion defined as domain $\Omega$ in the Cartesian coordinates $(x_1, x_2, x_3)$ given by

$$\frac{x_1^2}{a_1^2} + \frac{x_2^2}{a_2^2} + \frac{x_3^2}{a_3^2} \leq 1, \tag{0.1}$$

where the semiaxes are usually numbered so that $a_1 \geq a_2 \geq a_3$.

The parameter plays a crucial role in the Eshelby solution evaluation is $\lambda$, which is equal to the largest positive root of the equation:

$$\frac{x_1^2}{a_1^2 + \lambda} + \frac{x_2^2}{a_2^2 + \lambda} + \frac{x_3^2}{a_3^2 + \lambda} = 1. \tag{0.2}$$

A positive root always exists for points in the exterior domain ($x \notin \Omega$) and becomes equal to zero on the domain boundary $\partial\Omega$, as is evident from its definition in , Eq. (6.20). A positive root does not exist for points inside the ellipsoid, and $\lambda = 0$ should be assumed for interior points. It is convenient to think of the parameter $\lambda$ as a continuous single-valued real function of the field point x and the vector of ellipsoid semiaxes a:

$$\lambda = \lambda(\mathbf{x}, \mathbf{a}). \tag{0.3}$$

The elastic strain $e_{ij}$ within the inclusion that arises because of the introduction of eigenstrain $\varepsilon_{kl}^*$ is the result of multiplication by the *constant* tensor:

$$e_{ij} = S_{ijkl}\varepsilon_{kl}^*, \quad \mathbf{x} \in \Omega \tag{0.4}$$

where the interior Eshelby tensor $S_{ijkl} = S_{jikl} = S_{ijlk}$ possesses minor (but not major) indicial symmetries and depends only on the ellipsoid semiaxes and material's Poisson's ratio.

In contrast, the exterior elastic field depends on the coordinate x:

$$e_{ij}(\mathbf{x}) = D_{ijkl}(\mathbf{x})\varepsilon_{kl}^*, \quad \mathbf{x} \notin \Omega. \tag{0.5}$$

The interior Eshelby solution for an isotropic material is given by (Mura, 1987):

$$
\begin{aligned}
S_{1111} &= \frac{3}{8\pi(1-\nu)}a_1^2 I_{11} + \frac{(1-2\nu)}{8\pi(1-\nu)}I_1, \\[2mm]
S_{1122} &= \frac{1}{8\pi(1-\nu)}a_2^2 I_{12} - \frac{(1-2\nu)}{8\pi(1-\nu)}I_1, \\[2mm]
S_{1133} &= \frac{1}{8\pi(1-\nu)}a_3^2 I_{13} - \frac{(1-2\nu)}{8\pi(1-\nu)}I_1, \\[2mm]
S_{1212} &= \frac{a_1^2 + a_2^2}{16\pi(1-\nu)}I_{12} - \frac{(1-2\nu)}{16\pi(1-\nu)(I_1 + I_2)}.
\end{aligned}
\tag{0.6}
$$

The exterior Eshelby solution for an isotropic material is given by (Mura, 1987):

$$
\begin{aligned}
8\pi(1-\nu)D_{ijkl}(\mathbf{x}) &= 8\pi(1-\nu)D_{ijkl} + 2\nu\delta_{kl}x_i I_{I,j} \\
&+ (1-\nu)\big(\delta_{il}x_k I_{K,j} + \delta_{jl}x_k I_{K,i} + \delta_{ik}x_l I_{L,j} + \delta_{jk}x_l I_{L,i}\big) - \delta_{ij}x_k \big(I_K - a_I^2 I_{KI}\big)_l \\
&- \big(\delta_{ik}x_j + \delta_{jk}x_i\big)\big(I_J - a_I^2 I_{IJ}\big)_l - \big(\delta_{il}x_j + \delta_{jl}x_i\big)\big(I_J - a_I^2 I_{IJ}\big)_k - x_i x_j \big(I_J - a_I^2 I_{IJ}\big)_{lk},
\end{aligned}
\tag{0.7}
$$

where, following Mura (1987) the notation is adopted that summation from 1 to 3 is implied for lower case repeated indices, whereas upper case indices are not summed, but take on the same values as the corresponding lower case ones. All integral expressions $I_i$, $I_{ij}$ have implied dependence on parameter $\lambda$ and are given by

$$
\begin{aligned}
I_1 &= 2\pi a_1 a_2 a_3 \int_\lambda^\infty \frac{ds}{\left(a_1^2 + s\right)\Delta(s)}, \\[2mm]
I_{11} &= 2\pi a_1 a_2 a_3 \int_\lambda^\infty \frac{ds}{\left(a_1^2 + s\right)^2 \Delta(s)}, \\[2mm]
I_{12} &= 2\pi a_1 a_2 a_3 \int_\lambda^\infty \frac{ds}{\left(a_1^2 + s\right)\left(a_2^2 + s\right)\Delta(s)},
\end{aligned}
\tag{0.8}
$$

where $\Delta(s) = \left(a_1^2 + s\right)^{1/2}\left(a_2^2 + s\right)^{1/2}\left(a_3^2 + s\right)^{1/2}$, and the integrals for all other indices are obtained by cyclic permutations. The integrals are expressed in terms of the elliptic functions in the form:

$$
\begin{aligned}
I_1 &= \frac{4\pi a_1 a_2 a_3}{\left(a_1^2 - a_2^2\right)\left(a_1^2 - a_3^2\right)^{1/2}}[F(\theta, k) - E(\theta, k)], \\[3mm]
I_3 &= \frac{4\pi a_1 a_2 a_3}{\left(a_2^2 - a_3^2\right)\left(a_1^2 - a_3^2\right)^{1/2}}\left[\frac{\left(a_2^2 + \lambda\right)\left(a_1^2 - a_3^2\right)^{1/2}}{\prod_m\left(a_m^2 + \lambda\right)^{1/2}} - E(\theta, k)\right],
\end{aligned}
\tag{0.9}
$$

where the dependence of $\theta$ on $\lambda$ is implied, and $\lambda = 0$ should be used in the expressions related to the interior domain. The elliptic functions $F(\theta, k)$ and $E(\theta, k)$ have their usual definitions:

$$F(\theta, k) = \int_0^\theta \left(1 - k^2 \sin^2 t\right)^{-1/2} dt, \quad E(\theta, k) = \int_0^\theta \left(1 - k^2 \sin^2 t\right)^{-1/2} dt,$$

(0.10)

where parameters $\theta$, $k$ are:

$$\theta(\lambda) = \arcsin\left(\frac{a_1^2 - a_3^2}{a_1^2 + \lambda}\right)^{1/2}, \quad k = \left(\frac{a_1^2 - a_2^2}{a_1^2 - a_3^2}\right)^{1/2}.$$

(0.11)

The expressions for other $I$-integrals and their derivatives required for the evaluation of the exterior Eshelby tensor are written as follows:

$$I_2 = \frac{4\pi a_1 a_2 a_3}{\prod_m \left(a_m^2 + \lambda\right)^{1/2}} - I_1 - I_3,$$

$$I_{ij} = -\frac{I_i - I_j}{a_i^2 - a_j^2}, i \neq j,$$

(0.12)

$$I_{ii} = \frac{4\pi a_1 a_2 a_3}{3\left(a_i^2 + \lambda\right)\prod_m \left(a_m^2 + \lambda\right)^{1/2}} - \frac{\sum_k I_{ik}}{3}, i = j.$$

The derivatives are given by:

$$I_{i,j} = \frac{-2\pi a_1 a_2 a_3}{\left(a_i^2 + \lambda\right)\prod_m \left(a_m^2 + \lambda\right)^{1/2}}\lambda_j, \quad I_{ij,k} = \frac{-2\pi a_1 a_2 a_3}{\left(a_i^2 + \lambda\right)\left(a_j^2 + \lambda\right)\prod_m \left(a_m^2 + \lambda\right)^{1/2}}\lambda_k,$$

$$I_{ij,k} = \frac{-2\pi a_1 a_2 a_3}{\left(a_i^2 + \lambda\right)\prod_m \left(a_m^2 + \lambda\right)^{1/2}}\left(\lambda_{jk} - \left(\frac{1}{a_i^2 + \lambda} + \frac{1}{2}\sum_n \frac{1}{a_n^2 + \lambda}\right)\lambda_j\lambda_k\right),$$

$$I_{ij,kl} = \frac{-2\pi a_1 a_2 a_3}{\left(a_i^2 + \lambda\right)\left(a_j^2 + \lambda\right)\prod_m \left(a_m^2 + \lambda\right)^{1/2}}\left(\lambda_{kl} - \left(\frac{1}{a_i^2 + \lambda} + \frac{1}{a_j^2 + \lambda} + \frac{1}{2}\sum_n \frac{1}{a_n^2 + \lambda}\right)\lambda_k\lambda_l\right).$$

(0.13)

Finally, the aforementioned derivatives of $\lambda$ are written according to definition as:

$$\lambda_i = \frac{F_i}{C}, \quad \lambda_{ij} = \frac{F_{i,j} - \lambda_i C_j}{C},$$

(0.14)

where

$$F_i = \frac{2x_i}{a_i^2 + \lambda}, \quad C = \frac{x_m x_m}{\left(a_M^2 + \lambda\right)^2}.$$

(0.15)

Calculation of the exterior strain fields in particular using the expressions given earlier requires a combination of analytical manipulation and numerical evaluation. This can be accomplished using Matlab, but can also be done particularly efficiently using Mathematica. For example, the implementation of the function $\lambda = \lambda(x, a)$ is accomplished in one line as follows:

$$\text{lambda}\left[x\_, a\_\right] := \text{Max}\left[\lambda /. \text{NSolve}\left[\text{Plus@@}\left(x^\wedge 2/\left(a^\wedge 2 + \lambda\right)\right) = = 1, \{\lambda\}\right]\right];$$

(0.16)

The implementation can proceed first by defining the elliptic parameter $k$, which is independent of $\lambda$. Next, parameters $\theta$, $C$ and $F_i$, which depend on $\lambda$, can be defined as functions, and the necessary integrals also written in function form open to evaluation.

# Appendix B

На холмах Грузии лежит ночная мгла;
Шумит Арагва предо мною.
Мне грустно и легко; печаль моя светла;
Печаль моя полна тобою,
Тобой, одной тобой... Унынья моего
Ничто не мучит, не тревожит,
И сердце вновь горит и любит — оттого,
Что не любить оно не может.

The hills of Georgia lie in the nightly mist;
Aragva flows, fast and loud.
My soul is light and sad; my sorrow — soft and sweet;
My sorrow is imbued with thou.
With thou, and thou alone… My melancholy dreams
Are calm and undisturbed, for now
My heart beats once again with love — because it seems,
That otherwise it knows not how.

*AMK*

# Appendix C

*Marina Tsvetaeva. Monologue*

Уж сколько их упало в эту бездну,
Развёрстую вдали,
Настанет день, когда и я исчезну
С поверхности Земли.
Застынет всё, что пело и боролось,
Сияло и рвалось,
И зелень глаз моих, и нежный голос,
И золото волос.

И будет жизнь с её насущным хлебом,
С забывчивостью дня,
И будет всё, как будто бы под небом
И не было меня.
Изменчивой, как дети, в каждой мине
И так недолго злой,
Любившей час, когда дрова в камине
Становятся золой.

Виолончель и кавалькады в чаще,
И колокол в селе,
Меня такой живой и настоящей
На ласковой земле.
К вам всем что мне, ни в чём не знавшей меры,
Чужие и свои,
Я обращаюсь с требованьем веры
И с просьбой о любви.

How many souls have crossed that line already
Drawn somewhere ahead…
The day will come, and I will too be heading
Beyond that final net…
And all my passions, struggles and desires
Will vanish in a spin:
My gentle voice, my eye's soft fire,
The velvet of my skin…

The daily thread will run along unsevered
Through fabric of your life,
And it might seem to you as if I never
Walked under this blue sky.
Forgetting how I cherished sombre hour
When firewood turns to ash…
Forgetting how I always found the power
To face the world afresh…

How I adored the pensive voice of cello
The rustle of old leaves
That autumn paints so purple and so yellow
By sweeping of its sleeves…
I'll leave behind all of my heart's belongings
Below and above
I'll pass to you all my beliefs and longings
And a request for love.

За то, что мне прямая неизбежность
Прощение обид,
За всю мою безудержную нежность
И слишком гордый вид.
За быстроту стремительных событий,
За правду, за игру,
Послушайте ещё - меня любите
За то, что я умру.

К вам всем что мне, ни в чём не знавшей меры,
Чужие и свои,
Я обращаюсь с требованьем веры
И с просьбой о любви.

Forgive me if I have been too forgiving
Of others' hurtful words…
For tenderness for which I knew no limits
And frequent proud thoughts.
For fast and furious way I burnt my candle
For having never lied…
For all the things I know you will remember
When I have died…

I leave with you all of my heart's belongings
Below and above.
I leave with you all my beliefs and longings
And a request for love.

*AMK, for Katerina Reed-Tsocha*

# Bibliography

Abbey, B., Zhang, S.Y., Vorster, W., Korsunsky, A.M., 2012. Reconstruction of axisymmetric strain distributions via neutron strain tomography. Nucl. Instrum. Methods Phys. Res. B 270, 28–35.

Antunes, J.M., Fernandes, J.V., Sakharova, N.A., Oliveira, M.C., Menezes, L.F., 2007. On the determination of the Young's modulus of thin films using indentation tests. Int. J. Solids Struct. 44, 8313–8334.

Antunes, A.B., Ceretti, M., Paulus, W., Roisnel, T., Gil, V., Moure, C., Peña, O., 2008. Magnetic domains and anisotropy in single crystals of Er(Co,Mn)O$_3$. J. Magn. Magn. Mater. 320, 69–72.

Baimpas, N., Le Bourhis, E., Eve, S., 2013. Stress evaluation in thin films: micro-focus synchrotron X-ray diffraction combined with focused ion beam patterning for do evaluation. Thin Solid Films 549, 245–250.

Baimpas, N., Xie, M., Song, X., Hofmann, F., Abbey, B., Marrow, J., Mostafavi, M., Mi, J., Korsunsky, A.M., 2014a. Int. J. Comput. Methods 11, 1343006, 18 pages. http://dx.doi.org/10.1142/S0219876213430068.

Baimpas, N., Lunt, A.J.G., Dolbnya, I.P., 2014b. Nano-scale mapping of lattice strain and orientation inside carbon core SiC fibres by synchrotron X-ray diffraction. Carbon 79, 85–92.

Bemporad, E., Brisotto, M., Depero, L.E., Gelfi, M., Korsunsky, A.M., Lunt, A.J.G., Sebastiani, M., 2014. A critical comparison between XRD and FIB residual stress measurement techniques in thin films. Thin Solid Films 572, 224–231.

Bhavsar, S.N., Aravindan, S., Rao, P.V., 2012. Experimental investigation of redeposition during focused ion beam milling of high speed steel. Precis. Eng. 36, 408–413.

Blaber, J., Adair, B., Antoniou, A., 2015. Ncorr: open-source 2D digital image correlation Matlab software. Exp. Mech. 1–18.

Boyce, B.L., Chen, X., Hutchinson, J.W., 2001. The residual stress state due to a spherical hard-body impact. Mech. Mater. 33, 441–454.

Bravman, J.C., Sinclair, R., 1984. The preparation of cross-section specimens for transmission electron microscopy. J. Electron Microsc. Tech. 1, 53–61.

Carpinteri, A., 2002. Structural Mechanics: A Unified Approach. Taylor & Francis, Abingdon.

Chang, J.-Y., Yu, G.-P., Huang, J.-H., 2009. Determination of Young's modulus and Poisson's ratio of thin films by combining sin2ψ X-ray diffraction and laser curvature methods. Thin Solid Films 517, 6759–6766.

Chaves, J.M., Florêncio, O., Silva Jr., P.S., Marques, P.W.B., Afonso, C.R.M., 2015. Influence of phase transformations on dynamical elastic modulus and anelasticity of beta Ti–Nb–Fe alloys for biomedical applications. J. Mech. Behav. Biomed. Mater. 46, 184–196.

Cheng, W., Finnie, I., 2007. Residual Stress Measurement and the Slitting Method. Springer, New York.

Cho, S., Chasiotis, I., Friedmann, T.A., Sullivan, J.P., 2005. Young's modulus, Poisson's ratio and failure properties of tetrahedral amorphous diamond-like carbon for MEMS devices. J. Micromech. Microeng. 15, 728–735.

Chu, T., Ranson, W., Sutton, M., 1985. Applications of digital-image-correlation techniques to experimental mechanics. Exp. Mech. 25, 232–244.

Cleveringa, H.H.M., Van der Giessen, E., Needleman, A., 2000. A discrete dislocation analysis of mode I crack growth. J. Mech. Phys. Solids 48 (6–7), 1133–1157. http://dx.doi.org/10.1016/S0022-5096(99)00076-9.

Clément, L., Pantel, R., Kwakman, L.T., 2004. Strain measurements by convergent-beam electron diffraction: the importance of stress relaxation in lamella preparations. Appl. Phys. Lett. 85, 651–653.

Correlative Raman-SEM microscopy (RISE microscopy) in life sciences. Spectroscopy, February 2015. WITec GmbH, Ulm.

Davies, R.A., 2006. New batch-processing data-reduction application for X-ray diffraction data. J. Appl. Crystallogr. 39, 267–272.

Den Hartog, J.P., 1949. Strength of Materials. Dover, New York.

Eberl, C., Digital Image Correlation and Tracking. http://www.mathworks.co.uk/matlabcentral/fileexchange/12413-digital-image-correlation-and-tracking.

Eberl, C., Thompson, R., Gianola, D., Sharpe Jr., W., Hemker, K., 2006. Digital Image Correlation and Tracking. MatLabCentral, Mathworks File Exchange Server. FileID − 12413.

Eshelby, J.D., 1957. The determination of the elastic field of an ellipsoidal inclusion, and related problems. Proc. Roy. Soc. A. http://dx.doi.org/10.1098/rspa.1957.0133.

Farebrother, R.W., 1988. Linear Least Squares Computations. Taylor & Francis.

Flannery, C.M., Murray, C., Streiter, I., Schulz, S.E., 2001. Characterization of thin-filmaerogel porosity and stiffness with laser-generated surface acoustic waves. Thin Solid Films 388, 1–4.

Gelfi, M., Bemporad, E., Mariangela, B., 2014. A critical comparison between XRD and FIB residual stress measurement techniques in thin films. Thin Solid Films 572, 224–231.

Grogan, D.F., Zhao, T., Bovard, B.G., Macleod, H.A., 1992. Planarizing technique for ion-beam polishing of diamond films. Appl. Opt. 31, 1483–1487.

Gaucherin, G., Korsunsky, A.M., et al., 2009. Procedia Engineering, vol. 1.

Harris, G.L., 1995. Properties of Silicon Carbide. INSPEC, IET, Stevenage.

Henson, R.W., Reynolds, W.N., 1965. Lattice parameter changes in irradiated graphite. Carbon 3, 277–287.

Hild, F., Roux, S., 2012. Digital Image Correlation. Wiley-VCH, Weinheim.

Hills, D.A., Kelly, P.A., Dai, D.N., Korsunsky, A.M., 1996. Solution of Crack Problems: The Distributed Dislocation Technique. Springer. http://dx.doi.org/10.1007/978-94-015-8648-1.

Hoffmann, M., Birringer, R., 1995. Quantitative measurements of Young's modulus using the miniaturized disk-bend test. Mater. Sci. Eng. A 202, 18–25.

Huang, H., Dabiri, D., Gharib, M., 1997. On errors of digital particle image velocimetry. Meas. Sci. Technol. 8, 1427.

Huang, Y.J., et al., 2013. In situ study of the evolution of atomic strain of bulk metallic glass and its effects on shear band formation. Scr. Mater. 69, 207–210. http://dx.doi.org/10.1016/j.scriptamat.2013.03.016.

Hull, D., Bacon, D.J., 1984. Introduction to Dislocations. Pergamon, Oxford.

Jameson, E.C., 2001. Electrical Discharge Machining. Society of Manufacturing Engineers.

Jennett, N.M., Aldrich-Smith, G., Maxwell, A.S., 2004. Validated measurement of Young's modulus, Poisson ratio, and thickness for thin coatings by combining instrumented nanoindentation and acoustical measurements. J. Mater. Res. 19, 143–148.

Kalkman, A.J., Verbruggen, A.H., Janssen, G.C.A.M., 2003. High-temperature bulge-test setup for mechanical testing of free-standing thin films. Rev. Sci. Instrum. 74, 1383–1385.

Kang, K.J., Yao, N., He, M.Y., Evans, A.G., 2003. A method for in situ measurement of the residual stress in thin films by using the focused ion beam. Thin Solid Films 443, 71–77.

Kang, Y., Qiu, Y., Lei, Z., 2005. An application of Raman spectroscopy on the measurement of residual stress in porous silicon. Opt. Laser Eng. 43, 847–855.

Kassner, M.E., Nemat-Nasser, S., Suo, Z., Bao, G., Barbour, J.C., Brinson, L.C., Espinosa, H., Gao, H., Granick, S., Gumbsch, P., Kim, K.-S., Knauss, W., Kubin, L., Langer, J., Larson, B.C., Mahadevan, L., Majumdar, A., Torquato, S., van Swol, F., 2005. New directions in mechanics. Mech. Mater. 37 (2–3), 231–259.

Kyungmok, K., DPhil thesis, Oxford, 2005.

Keil, S., 1992. Experimental determination of residual stresses with the ring-core method and an on-line measuring system. Exp. Tech. 16, 17–24.

Keller, R., Geiss, R., 2012. Transmission EBSD from 10 nm domains in a scanning electron microscope. J. Microsc. 245, 245–251.

King, A., Johnson, G., Engelberg, D., 2008. Observations of intergranular stress corrosion cracking in a grain-mapped polycrystal. Science 321, 382–385.

Korsunsky, A.M., 2005a. On the modelling of residual stresses due to surface peening using eigenstrain distributions. J. Strain Anal. 40.

Korsunsky, A.M., 2005b. Residual elastic strain due to laser shock peening. J. Strain Anal. 40.

Korsunsky, A.M., 2006a. Variational eigenstrain analysis of synchrotron diffraction measurements of residual elastic strain in a bent titanium alloy bar. J. Mech. Mater. Struct. 1, 259–277.

Korsunsky, A.M., 2006b. Residual elastic strain due to laser shock peening: modelling by eigen strain distribution. J. Strain Anal. 41, 195–204. http://dx.doi.org/10.1243/03093247JSA141.

Korsunsky, A.M., Regino, G.M., July 2007. Residual elastic strains in autofrettaged tubes: variational analysis by the eigenstrain finite element method. J. Appl. Mech. 4 (74), 717–722.

Korsunsky, A.M., 2007. Residual elastic strains in autofrettaged tubes: elastic–ideally plastic model analysis. J. Eng. Mat. Technol. ASME 129, 77–81.

Korsunsky, A.M., Constantinescu, A., 2006. Work of indentation approach to the analysis of hardness and modulus of thin coatings. Mater. Sci. Eng. A Struct. 423, 28–35.

Korsunsky, A.M., Constantinescu, A., 2009. The influence of indenter bluntness on the apparent contact stiffness of thin coatings. Thin Solid Films 517, 4835–4844.

Korsunsky, A.M., Regino, G.M., Nowell, D., 2007. Variational eigenstrain analysis of residual stresses in a welded plate. Int. J. Solids Struct. 44 (13), 4574–4591. http://dx.doi.org/10.1016/j.ijsolstr.2006.11.037.

Korsunsky, A.M., Sebastiani, M., Bemporad, E., 2009. Focused ion beam ring drilling for residual stress evaluation. Mater. Lett. 63, 1961–1963.

Korsunsky, A.M., Sebastiani, M., Bemporad, E., 2010. Residual stress evaluation at the micrometer scale: analysis of thin coatings by FIB milling and digital image correlation. Surf. Coat. Technol. 205, 2393–2403.

Korsunsky, A.M., Sui, T., Salvati, E., George, E.P., Sebastiani, M., 2016. Experimental and modelling characterisation of residual stresses in cylindrical samples of rapidly cooled bulk metallic glass. Mater. Des. 104, 235–241.

Korsunsky, A.M., Withers, P.J., 1997. Plastic bending of a residually stressed beam. Int. J. Solids Struct. 34 (16), 1985−2002.

Krieger Lassen, N.C., Juul Jensen, D., Conradsen, K., 1992. Image processing procedures for analysis of electron back scattering patterns. Scanning Microsc. 6, 115−121.

Krottenthaler, M., Schmid, C., Schaufler, J., 2013. A simple method for residual stress measurements in thin films by means of focused ion beam milling and digital image correlation. Surf. Coat. Tech. 215, 247−252.

Landolt, D., 1987. Fundamental aspects of electropolishing. Electrochim. Acta 32, 1−11.

Liang, C., Prorok, B.C., 2007. Measuring the thin film elastic modulus with a magnetostrictive sensor. J. Micromech. Microeng. 17, 709−716.

Liu, D., Kyaw, S., Flewitt, P., 2014. Residual stresses in environmental and thermal barrier coatings on curved superalloy substrates: experimental measurements and modelling. Mat. Sci. Eng. A Struct. 606, 117−126.

Lunt, A.J.G., Korsunsky, A.M., 2014. Intragranular residual stress evaluation using the semi-destructive FIB-DIC ring-core drilling method. Adv. Mater. Res. 996, 8−13.

Lunt, A., Xie, M., Baimpas, N., Zhang, S., Kabra, S., Kelleher, J., Neo, T., Korsunsky, A., 2014. Calculations of single crystal elastic constants for yttria partially stabilised zirconia from powder diffraction data. J. Appl. Phys. 116, 053509.

Lunt, A.J.G., Korsunsky, A.M., 2015. A review of micro-scale focused ion beam milling and digital image correlation analysis for residual stress evaluation and error estimation. Surf. Coatings Technol. 283, 373−388. http://dx.doi.org/10.1016/j.surfcoat.2015.10.049.

Lunt, J.G., Baimpas, N., Salvati, E., Dolbnya, I., Sui, T., Ying, S., Zhang, H., Kleppe, A., Dluhos, J., Korsunsky, A.M., 2015. A state-of-the-art review of micron scale spatially resolved residual stress analysis by ring-core milling and other techniques. J. Strain Anal. Eng. 50, 426−444. http://dx.doi.org/10.1177/0309324715596700.

Lunt, A.J.G., Salvati, E., Ma, L., Dolbnya, I.P., Neo, T.K., Korsunsky, A.M., 2016. Full in-plane strain tensor analysis using the microscale ring-core FIB milling and DIC approach. J. Mech. Phys. Solids 94, 47−67.

Macherauch, E., 1987. Origin, measurement and evaluation of residual stresses. In: Macherauch, E., Hauk, V. (Eds.), Residual Stresses in Science and Technology, Proceedings of the International Conference on Residual Stresses, 1986 Garmisch-partenkirchen (FRO). DOM Informationsgesellschaft, Oberursel, p. 3.

Mansilla, C., Martínez-Martínez, D., Ocelík, V., De Hosson, J.T.M., 2015. On the determination of local residual stress gradients by the slit milling method. J. Mater. Sci. 50, 3646−3655.

Masing, G., 1923. Zur Hayn'ischen Theorie der Verfestigung der Metalle durch verborgene elastische Spannungen. Wiss. Veroff. Siemens Konz. 3, 231−239.

Martins, R.V., Ohms, C., Decroos, K., 2010. Full 3D spatially resolved mapping of residual strain in a 316L austenitic stainless steel weld specimen. Mat. Sci. Eng. A Struct. 527, 4779−4787.

Marturi, N., Dembélé, S., Piat, N., 2013. Fast image drift compensation in scanning electron microscope using image registration. Trans. Autom. Sci. Eng. IEEE 807−812.

McCarthy, J., Pei, Z., Becker, M., 2000. FIB micromachined submicron thickness cantilevers for the study of thin film properties. Thin Solid Films 358, 146−151.

McCormick, N., Lord, J., 2010. Digital image correlation. Mater. Today 13, 52−54.

Mindlin, R.D., Chen, D.H., 1950. Nuclei of strain in the semi-infinite solid. J. Appl. Phys. 21, 926−930.

Mohanty, G., Wheeler, J.M., Raghavan, R., Wehrs, J., Hasegawa, M., Mischler, S., Philippe, L., Michler, J., 2014. Elevated temperature, strain rate jump microcompression of nanocrystalline nickel. Philos. Mag. 1−18.

Moncrieff, D., Robinson, V., Harris, L., 1978. Charge neutralisation of insulating surfaces in the SEM by gas ionisation. J. Phys. D Appl. Phys. 11, 2315.

Mortazavian, S., Fatemi, A., 2015. Effects of fiber orientation and anisotropy on tensile strength and elastic modulus of short fiber reinforced polymer composites. Compos Part B Eng. 72, 116–129.

Mukherji, D., Gilles, R., Barbier, B., 2003. Lattice misfit measurement in Inconel 706 containing coherent $\gamma'$ and $\gamma''$ precipitates. Scr. Mater. 48, 333–339.

Mura, T., 1987. Micromechanics of Defects in Solids, second ed. Springer, 588p. http://dx.doi.org/10.1007/978-94-009-3489-4.

Nava, R., 1998. Evaluation of the high-temperature pressure derivative of the Grüneisen constant from the temperature variation of the elastic moduli. J. Phys. Chem. Solids 59, 1537–1539.

Nelson, D.V., 2010. Residual stress determination by hole drilling combined with optical methods. Exp. Mech. 50, 145–158.

Newbury, D.E., 2002. X-ray microanalysis in the variable pressure (environmental) scanning electron microscope. J. Res. Natl. Inst. Stand. 107, 567–604.

Nicola, L., Van der Giessen, E., Needleman, A., 2003. Discrete dislocation analysis of size effects in thin films. J. Appl. Phys. 93 (10), 5920–5928.

Noyan, I.C., Cohen, J.B., 2013. Residual Stress: Measurement by Diffraction and Interpretation. Springer-Verlag.

Ogden, R.W., 1984. Non-linear Elastic Deformations. Ellis Horwood, Chichester.

Oka, Y., Kirinuki, M., Nishimura, Y., 2004. Measurement of residual stress in DLC films prepared by plasma-based ion implantation and deposition. Surf. Coat. Tech. 186, 141–145.

Okai, N., Sohda, Y., 2012. Study on image drift induced by charging during observation by scanning electron microscope. Jpn. J. Appl. Phys. 51, 6–11.

Pan, B., Xie, H., Guo, Z., Hua, T., 2007. Full-field strain measurement using a two dimensional Savitzky–Golay digital differentiator in digital image correlation. Opt. Eng. 46 (033601–033601–033610).

Pan, B., Xie, H., Wang, Z., Qian, K., Wang, Z., 2008. Study on subset size selection in digital image correlation for speckle patterns. Opt. Express 16, 7037–7048.

Pestka, K.A., Maynard, J.D., Gao, D., Carraro, C., 2008. Measurement of the elastic constants of a columnar SiC thin film. Phys. Rev. Lett. 100, 055503.

Poulsen, H., et al., 2005. Measuring strain distributions in amorphous materials. Nat. Mater. 4, 33–36. http://dx.doi.org/10.1038/nmat1266.

Prime, M.B., 1999. Residual stress measurement by successive extension of a slot: the crack compliance method. Appl. Mech. Rev. 52, 75–96.

Qiu, S., Clausen, B., Padula Ii, S.A., Noebe, R.D., Vaidyanathan, R., 2011. On elastic moduli and elastic anisotropy in polycrystalline martensitic NiTi. Acta Mater. 59, 5055–5066.

Read, D.T., Cheng, Y.-W., Keller, R.R., McColskey, J.D., 2001. Tensile properties of freestanding aluminum thin films. Scr. Mater. 45, 583–589.

Renault, P.O., Badawi, K.F., Bimbault, L., Goudeau, P., Elkaïm, E., Lauriat, J.P., 1998. Poisson's ratio measurement in tungsten thin films combining an X-ray diffractometer with in situ tensile tester. Appl. Phys. Lett. 73, 1952–1954.

Roberts, O., Lunt, A.J.G., Ying, S., July 2–4, 2014. A study of phase transformation at the surface of a zirconia ceramic. In: Proceedings of the World Congress on Engineering, vol. 2. Newswood and International Association of Engineers, London, Hong Kong.

Sabate, N., Vogel, D., Gollhardt, A., 2006a. Measurement of residual stress by slot milling with focused ion-beam equipment. J. Micromech. Microeng. 16, 254–259.

Sabaté, N., Vogel, D., Gollhardt, A., 2006b. Measurement of residual stresses in micromachined structures in a microregion. Appl. Phys. Lett. 88, 071910.

Sabaté, N., Vogel, D., Gollhardt, A., Keller, J., Cané, C., Gràcia, I., Morante, J.R., Michel, B., 2007a. Residual stress measurement on a MEMS structure with high-spatial resolution. J. Microelectromech. Syst. 16, 365–372.

Sabaté, N., Vogel, D., Keller, J., 2007b. FIB-based technique for stress characterization on thin films for reliability purposes. Microelectron. Eng. 84, 1783–1787.

Sahoo, B., Satpathy, R.K., Prasad, K., 2013. Effect of shot peening on low cycle fatigue life of compressor disc of a typical fighter class aero-engine. Proc. Eng. 55, 144–148.

Salvati, E., Sui, T., Ying, S., 2014. On the accuracy of residual stress evaluation from focused ion beam DIC (FIB-DIC) ring-core milling experiments. In: Proceedings of the 5th International Conference on Nanotechnology: Fundamentals and Applications Prague August 11–13. Avestia, Ottawa, ON, p. 265.

Salvati, E., Sui, T., Ying, S., Lunt, A.J., Korsunsky, A.M., 2014a. On the accuracy of residual stress evaluation from Focused Ion Beam DIC (FIB-DIC) ring-core milling experiments. Proc. ICNFA'15 265.

Schajer, G.S., 1988. Measurement of non-uniform residual stresses using the hole-drilling method. J. Eng. Mater. Technol. 110 (4), Part I: 338–343, Part II: 344–349.

Schajer, G., Altus, E., 1996. Stress calculation error analysis for incremental hole-drilling residual stress measurements. J. Eng. Mater. Technol. 118, 120–126.

Sebastiani, M., Eberl, C., Bemporad, E., Pharr, G.M., 2011. Depth-resolved residual stress analysis of thin coatings by a new FIB–DIC method. Mater. Sci. Eng. A 528, 7901–7908.

Sebastiani, M., Eberl, C., Bemporad, E., Korsunsky, A.M., Nix, W.D., Carassiti, F., 2014. Focused ion beam four-slot milling for Poisson's ratio and residual stress evaluation at the micron scale. Surf. Coat. Technol. 251, 151–161.

Shan, X., Xiao, X., Liu, Y., 2011. Determination of Young's modulus and Poisson's ratio of nano-porous low-k thin film by laser-generated surface acoustic waves. Adv. Sci. Lett. 4, 1230–1234.

Sharpe Jr., W.N., Yuan, B., Edwards, R.L., 1997. A new technique for measuring the mechanical properties of thin films. J. Microelectromech. Syst. 6, 193–198.

Sheng, X-f, Xia, Q-x, Cheng, X-q, 2012. Residual stress field induced by shot peening based on random-shots for 7075 aluminum alloy. T Nonferr. Metal. Soc. 22 (Suppl. 2), 261–267.

Slama, C., Abdellaoui, M., 2000. Structural characterization of the aged Inconel 718. J. Alloy Compd. 306, 277–284.

Snigireva, I., Snigirev, A., 2006. X-ray microanalytical techniques based on synchrotron radiation. J. Environ. Monit. 8, 33–42.

Song, X., Yeap, K.B., Zhu, J., 2011. Residual stress measurement in thin films using the semi-destructive ring-core drilling method using Focused Ion Beam. Proc. Eng. 10, 2190–2195.

Song, X., Liu, W., Belnoue, J., 2012a. An eigen strain-based finite element model and the evolution of shot peening residual stresses during fatigue of GW103 magnesium alloy. Int. J. Fatigue 42, 284–295.

Song, X., Yeap, K.B., Zhu, J., Belnoue, J., Sebastiani, M., Bemporad, E., Zeng, K.Y., Korsunsky, A.M., 2012b. Residual stress measurement in thin films at sub-micron scale using Focused Ion Beam milling and imaging. Thin Solid Films 520, 2073–2076.

Suresh, K.S., Lahiri, D., Agarwal, A., Suwas, S., 2015. Microstructure dependent elasticmodulus variation in NiTi shape memory alloy. J. Alloys Compd. 633, 71–74.

Sutton, M.A., Orteu, J.J., Schreier, H., 2009. Image Correlation for Shape, Motion and Deformation Measurements: Basic Concepts, Theory and Applications. Springer Science & Business Media.

Suzuki, E., 2002. High-resolution scanning electron microscopy of immunogold-labelled cells by the use of thin plasma coating of osmium. J. Microsc. 208, 153–157.

Suzuki, S., 2013. Features of transmission EBSD and its application. JOM 65, 1254–1263.

Timoshenko, S., Goodier, J.N., 1951. Theory of Elasticity.

Timoshenko, S.P., Goodier, J.N., 1965. Strength of Materials. Prentice Hall, New York.

Tomioka, Y., Yuki, N., 2004. Bend stiffness of copper and copper alloy foils. J. Mater. Process. Technol. 146, 228–233.

Torres, M.A.S., Voorwald, H.J.C., 2002. An evaluation of shot peening, residual stress and stress relaxation on the fatigue life of AISI 4340 steel. Int. J. Fatigue 24, 877–886.

Tricoteaux, A., Duarte, G., Chicot, D., Le Bourhis, E., Bemporad, E., Lesage, J., 2010. Depthsensing indentation modeling for determination of elastic modulus of thin films. Mech. Mater. 42, 166–174.

Vaughan, G.B., Wright, J.P., Bytchkov, A., 2011. X-ray transfocators: focusing devices based on compound refractive lenses. J. Synchrotron Radiat. 18, 125–133.

Vigouroux, M., Delaye, V., Bernier, N., 2014. Strain mapping at the nanoscale using precession electron diffraction in transmission electron microscope with off axis camera. Appl. Phys. Lett. 105, 191906.

Vlassak, J.J., Nix, W.D., 1992. New bulge test technique for the determination of Young's modulus and Poisson's ratio of thin films. J. Mater. Res. 7, 3242–3249.

Vogel, D., Sabate, N., Gollhardt, A., February 26, 2006. FIB-based measurement of local residual stresses on microsystems. In: Proceedings of SPIE 6175, Testing, Reliability, and Application of Micro- and Nano-material Systems IV, San Diego, CA. SPIE, Bellingham, WA, pp. 617505–617508.

Wachtel, E., Lubomirsky, I., 2011. The elastic modulus of pure and doped ceria. Scr. Mater. 65, 112–117.

Wang, Y., Cuitiño, A.M., 2002. Full-field measurements of heterogeneous deformation patterns on polymeric foams using digital image correlation. Int. J. Solids Struct. 39, 3777–3796.

Wang, L., Rokhlin, S.I., 2002. Recursive asymptotic stiffness matrix method for analysis of surface acoustic wave devices on layered piezoelectric media. Appl. Phys. Lett. 81, 4049–4051.

Wang, Q., Ozaki, K., Ishikawa, H., 2006. Indentation method to measure the residual stress induced by ion implantation. Nucl. Instr. Meth Phys. Res. B 242, 88–92.

Watanabe, Y., Hasegawa, N., Matsumura, Y., 1995. Simulation of residual stress distribution on shot peening. J. Soc. Mater. Sci. 44, 110–116.

Wilkinson, A.J., Britton, T.B., 2012. Strains, planes, and EBSD in materials science. Mater. Today 15, 366–376.

Wilkinson, A.J., Meaden, G., Dingley, D.J., 2006. High-resolution elastic strain measurement from electron backscatter diffraction patterns: new levels of sensitivity. Ultramicroscopy 106, 307–313.

Winiarski, B., Withers, P.J., 2010. Mapping residual stress profiles at the micron scale using FIB micro-hole drilling. Appl. Mech. Mater. 24, 267–272.

Winiarski, B., Withers, P.J., 2012. Micron-scale residual stress measurement by micro-hole drilling and digital image correlation. Exp. Mech. 52, 417–428.

Winiarski, B., Withers, P.J., 2015. Novel implementations of relaxation methods for measuring residual stresses at the micron scale. J. Strain Anal. Eng. 50, 412–425. http://dx.doi.org/10.1177/0309324715590957.

Winiarski, B., Langford, R., Tian, J., Yokoyama, Y., Liaw, P., Withers, P., 2010. Mapping residual stress distributions at the micron scale in amorphous materials. Metall. Mater. Trans. A 41, 1743–1751.

Winiarski, B., Gholinia, A., Tian, J., Yokoyama, Y., Liaw, P.K., Withers, P.J., 2012a. Submicronscale depth profiling of residual stress in amorphous materials by incremental focused ion beam slotting. Acta Mater. 60, 2337—2349.

Winiarski, B., Schajer, G., Withers, P., 2012b. Surface decoration for improving the accuracy of displacement measurements by digital image correlation in SEM. Exp. Mech. 52, 793—804.

Yaofeng, S., Pang, J.H., 2007. Study of optimal subset size in digital image correlation of speckle pattern images. Opt. Lasers Eng. 45, 967—974.

Ye, J., Shimizu, S., Sato, S., Kojima, N., Noro, J., 2006. Bidirectional thermal expansion measurement for evaluating Poisson's ratio of thin films. Appl. Phys. Lett. 89, 031913.

Yu, H.Y., 2009. Variation of elastic modulus during plastic deformation and its influence on spring back. Mater. Des. 30, 846—850.

Zhang, S.Y., Vorster, W., Jun, T.S., 2008a. High energy white beam x-ray diffraction studies of residual strains in engineering components. In: Korsunsky, A.M. (Ed.), World Congress on Engineering 2007. Assoc Engineers, London, England. July 02—04, 2007.

Zhang, S.Y., Schlipf, J., Korsunsky, A.M., 2008b. Analysis of residual stresses around "dimpled" cold-expanded holes in aluminium alloy plates. Mater. Sci. Forum 571—572, 295—300.

Zhao, W., Seshadri, R., Dubey, R.N., 2003. On thick-walled cylinder under internal pressure. J. Press Vess T ASME 125, 267—273.

Zhu, W.L., Zhu, J.L., Nishino, S., 2006. Spatially resolved Raman spectroscopy evaluation of residual stresses in 3C-SiC layer deposited on Si substrates with different crystallographic orientations. Appl. Surf. Sci. 252, 2346—2354.

## FURTHER READING

ASM Engineered Materials Handbook. 1988 vol. 2, Engineering Plastics. ASM International, Metals Park, Ohio.

Bever, M.B. (Ed.), 1986. Encyclopedia of Materials Science and Engineering. Pergamon, Oxford.

Brandes, E.A., Brook, G.B., Smithells, C.J. (Eds.), 1998. Smithells Metals Reference Book. Butterworth-Heinemann, Oxford.

Carpinteri, A., 2002. Structural Mechanics: A Unified Approach. Taylor & Francis, Abingdon.

Cheng, W., Finnie, I., 2007. Residual Stress Measurement and the Slitting Method. Springer, New York.

Crawford, R.J., 1998. Plastics Engineering. Butterworth-Heinemann, Oxford.

Harris, G.L., 1995. Properties of Silicon Carbide. INSPEC, IET, Stevenage.

Kockelmann, W., Frei, G., Lehmann, E.H., Vontobel, P., Santisteban, J.R., 2007. Energy-selective neutron transmission imaging at a pulsed source. Nucl. Instrum. Methods Phys. Res. A 578, 421—434. http://dx.doi.org/10.1016/j.nima.2007.05.207.

Landau, L.D., Lifshitz, E.M., 1970. Theory of Elasticity. Pergamon, Oxford.

Lunt, A.J.G., Mohanty, G., Ying, S., Dluhos, J., Sui, T., Neo, T.K., Michler, J., Korsunsky, A.M., 2015. A correlative microscopy study of the zirconiaeporcelain interface in dental prostheses: TEM, EDS and micro-pillar compression. Thin Solid Films 596, 222—232. https://doi.org/10.1016/j.tsf.2015.07.070.

Nye, J.F., 2000. Physical Properties of Crystals. OUP, Oxford.

Pedersen, P., 1995. Simple transformations by proper contracted forms — can we change the usual practice? Comms. Num. Meth. Eng. 11, 821—829.

Rubin, I.I. (Ed.), 1990. Handbook of Plastic Materials and Technology. Wiley, New York.

Slama, C., Abdellaoui, M., 2000. Structural characterization of the aged Inconel 718. J. Alloy Compd. 306, 277–284.

Shackelford, J.F. (Ed.), 1994. CRC Materials Science and Engineering Handbook. CRC Press, Boca Raton.

Van der Giessen, E., Needleman, A., 2002. Micromechanics simulations of fracture. Ann. Rev. Mater. Res. 32, 141–162.

Van Krevelen, D.W., 1990. Properties of Polymers. Elsevier, Oxford.

# Index

Printed in the United States
By Bookmasters